W0080549

ISO 9000

Practices in Construction

ISO 9000

Practices in Construction

KB Rajoria ME
Former Engineer-in-Chief
PWD, Government of Delhi

Deepak Narayan ME MPhil
Former Engineer-in-Chief
PWD, Government of Delhi

Deepak Gupta BE, PGDM, MBA
Quality Management Expert

CBS

CBS Publishers & Distributors Pvt Ltd

New Delhi • Bengaluru • Chennai • Kochi • Kolkata • Mumbai
Hyderabad • Jharkhand • Nagpur • Patna • Puna • Uttarakhand

Disclaimer

Science and technology are constantly changing fields. New research and experience broaden the scope of information and knowledge. The authors have tried their best in giving information available to them while preparing the material for this book. Although, all efforts have been made to ensure optimum accuracy of the material, yet it is quite possible some errors might have been left uncorrected. The publisher, printer and the authors will not be held responsible for any inadvertent errors, omissions or inaccuracies.

ISO 9000

Practices in Construction

ISBN: 978-93-90709-33-5

Copyright © Authors and Publisher

First Edition: 2022

All rights reserved. No part of this book may be reproduced or transmitted in any form or by any means, electronic or mechanical, including photocopying, recording, or any information storage and retrieval system without permission, in writing, from the authors and the publisher.

Published by Satish Kumar Jain and produced by Varun Jain for

CBS Publishers & Distributors Pvt Ltd

4819/XI Prahlad Street, 24 Ansari Road, Daryaganj, New Delhi 110 002, India
Ph: 011-23289259, 23266861, 23266867   Website: www.cbspd.com
Fax: 011-23243014   e-mail: delhi@cbspd.com; cbspubs@airtelmail.in

Corporate Office: 204 FIE, Industrial Area, Patparganj, Delhi 110 092
Ph: 011-4934 4934   Fax: 011-4934 4935   e-mail: publishing@cbspd.com; publicity@cbspd.com

Branches

- **Bengaluru:** Seema House 2975, 17th Cross, K.R. Road, Banasankari 2nd Stage, Bengaluru 560 070, Karnataka
  Ph: +91-80-26771678/79   Fax: +91-80-26771680   e-mail: bangalore@cbspd.com
- **Chennai:** 7, Subbaraya Street, Shenoy Nagar, Chennai 600 030, Tamil Nadu
  Ph: +91-44-26680620, 26681266   Fax: +91-44-42032115   e-mail: chennai@cbspd.com
- **Kochi:** 42/1325, 1326 Power House Road, Opp. KSEB, Ernakulum, Kochi 682018, Kerala
  Ph: +91-484-4059061-65   Fax: +91-484-4059065   e-mail: kochi@cbspd.com
- **Kolkata:** 6/B, Ground Floor, Rameswar Shaw Road, Kolkata 700 014, West Bengal
  Ph: +91-33-22891126, 22891127, 22891128   e-mail: kolkata@cbspd.com
- **Mumbai:** PWD Shed, Gala No. 25/26, Ramchandra Bhatt Marg, Next to JJ Hospital, Gate No. 2,
  Opposite Union Bank of India, Noorbaug, Mumbai 400 009, Maharashtra, India
  Ph: +91-22-24902340/41   Fax: +91-22-24902342   e-mail: mumbai@cbspd.com

Representatives

- **Hyderabad** 0-9885175004
- **Pune** 0-9623451994
- **Jharkhand** 0-9811541605
- **Uttarakhand** 0-9716462459
- **Nagpur** 0-9421945513
- **Patna** 0-9334159340

Printed at Mudrak, Noida, UP, India

Foreword

The importance of quality in construction is well known, but despite the quality measures adopted, requisite level of quality is rarely achieved. The construction companies unfortunately have failed to institutionalize quality measures in their management systems and construction work. In an increasingly competitive market, a reliable quality management system is crucial to the successful completion of projects. The companies that establish ISO 9000 discipline will find it easier to accomplish their projects, leading to higher level of satisfaction and superior product quality.

Most government engineering departments have an in-house quality control or quality assurance unit. In Central Vigilance Commission there is also a technical audit unit (Chief Technical Examiner) to watch and monitor the quality of works in the country. Most public sector organizations also have an in-house vigilance unit to monitor their quality of works. Despite of all these measures, the quality of projects delivered generally leaves much to be desired. For a uniform quality product, the quality consciousness has to be ingrained into the ethos of the organization. The organizational structure and the project team has to be set up that it does not permit any sub-standard work to be executed. ISO 9000 series of standards propose to precisely address this issue.

The International Organization for Standardization (ISO) is a global organization that works to provide standardization across an array of products and services. Its main goal is to facilitate trade, but its focus is also on process improvement, safety, and quality in several areas. The goal of ISO 9000 is to embed a quality management system within an organization, increasing productivity, reducing unnecessary costs, and ensuring quality of processes and products. The use of "process approach" is a requirement for all ISO management system standards. ISO 9000 is a set of international standards on quality management developed to assist companies to document and maintain an efficient quality system. They are not specific to any one industry and can be applied across broad spectrum of companies. Having said this, why should one adopt these standards? Because this will result in continual improvement in operating process, reducing waste, increasing productivity, resulting in better marketing, and increasing customer satisfaction. Also, in bidding for international projects, ISO 9001 certification is often a prerequisite. Therefore, adopting of such standards widens the scope of business opportunities—enabling the organization to increase its market share and overall competitiveness. Whether you operate internationally or want to expand locally, ISO 9001 certification will demonstrate your commitment to quality.

The authors of this book have done a yeoman's job by writing a treatise on the ISO9000 series of standards explaining in detail as to how to go about in adopting these standards. The three authors have held high positions in engineering departments in the country and put together have nearly 125 years of experience in the field of engineering. Rarely is there such a combination of experience, mature farsightedness and administrative experience to produce such a useful document. The book is written in an easy-to-understand style in 8 chapters where the complex web of various quality related standards is explained thread bare. It is hoped that this book will help in better understanding of ISO 9000 practices and will help in raising quality standards of our construction works.

HS Dogra

Former Director General, CPWD, and
Chairman, Civil Engineering Division Council
Bureau of Indian Standards, GoI
Member of Indian Concrete Institute
Member of Indian Buildings Congress, and
Member of Indian Institute of Technical Arbitrators
Fellow, Indian Council of Abritration

Preface

Construction is the most important activity for growth and development of a country. By and large the construction industry is lagging behind the other industries in ensuring suitable quality standards. It is therefore, necessary to look into the quality management systems and take appropriate measures for quality improvements in construction industry. In order to deliver products and services at a consistently high levels of quality, it is considered necessary to adopt ISO 9000 practices in construction. In the Western world that is Europe and USA, most of the consultants and contractors have adopted ISO 9000 standards in their working. In order to move forward, it is necessary to ensure implementation of international quality standards particularly in major construction projects. Therefore, engineers, architects, planners, designers, contractors and suppliers, all will have to understand and implement ISO 9000 practices in their organization, to deliver products and services at desired levels of quality.

For a construction project, input is required from principal employer, consultants, civil contractor, services contractors, suppliers, subcontractors, etc. It is seen that all the agencies involved may or may not have ISO management system standard certification. The quality standards can be achieved for the project through combined effort of all concerned. Therefore, the principal employer has to ensure achieving this objective by training concerned persons of all stakeholders and developing systems and procedures according to requirements of ISO 9000 standards. The objective is to complete the project according to ISO 9000 standards and further develop maintenance services according to these standards.

The purpose of writing this book is to make Indian construction projects and products and services equitable to international standards of quality. This book will help consultants, planners, suppliers and contractors to move towards planning and implementing the ISO 9000 standards and helping in augmenting the performance, and improving the productivity and efficiency of the organization. ISO 9000 is a proven method of building a quality track record that will withstand the closest scrutiny even in the most competitive environment. ISO 9000 practices in construction enable construction professionals to implement quality standards and procedures precisely suited to their needs and responsibilities.

Engineering institutions, i.e. IITs, NITs, other engineering collages, polytechnics, etc. should also consider including in their curriculum, studies about ISO standards for engineering students. Some institutes have incorporated elementary introduction to ISO provisions in their engineering studies. However, it is considered desirable to have extensive coverage in syllabus for adopting ISO standards in construction. For postgraduate studies special courses can be introduced to cover this field of knowledge. India is a fast moving from a developing country to the status of a developed country. Dissipation of knowledge about ISO 9000 standards to different stakeholders connected with construction works, and implementation of ISO standards is a correct step in the direction of delivering products and services at a consistently high levels of quality.

KB Rajoria

Acknowledgements

This book is dedicated to Indian Buildings Congress (IBC), in view of long-standing association of the authors with this national organization promoting quality and sustainability in built environment. We express our sincere thanks to Shri Pradeep Mittal, President of Indian Buildings Congress and members of the Executive Committee for their support and encouragement in this endeavour.

It was considered desirable to get views and suggestions from some outstanding professionals of our country in this field. Accordingly we approached Shri P Krishnan, Faculty in Engineering Institutes and former Director General, Central PWD, Shri V Suresh, Chairman of Indian Green Buildings Congress and former Chairman and Managing Director, HUDCO, Shri Anil Jauhri, former CEO, National Accreditation Board for Certification Bodies, Quality Council of India and Shri ANSP Sastry, international consultant on management systems, specialist in ISO 9001 and former Director, BIS. We are grateful to them for their valuable suggestions which have been suitably incorporated in the book.

We express our gratitude to Shri KK Kapila, CMD, Intercontinental Consultants and Technocrats Pvt. Ltd, for providing technical support, in the writing of this book. We are also thankful to Shri Shobit Uppal, Deputy CMD of Ahluwalia Contracts (India) Ltd. We have visited their projects duly accompanied by Shri Amarjeet Singh, DGM, Ahluwalia Contracts (India) Ltd. We are also grateful to Shri Jagmohan Lal, former Additional DG, CPWD, for providing case study pertaining to construction project implemented by him as per ISO standards.

KB Rajoria
Deepak Narayan
Deepak Gupta

Contents

Introduction

In order to achieve the objective of implementing quality management systems according to ISO 9000 standards, it is necessary to have a comprehensive study of related standards and the background of development of these standards. The detailed study of ISO Standards related to quality and other standard, has been complied in this book which will help in achieving implementation of quality in construction. The construction industry does not have streamlined mandatory procedures, rules and regulations. It needs self-imposed and initiated policies in respect of quality management to be followed by the Government and the construction industry set up, to establish and implement ISO 9000 practices in construction.

The basic understanding of ISO management system standards and implementation will help construction industry in India to move forward to international levels in quality. The quality management principles will help to mitigate many of the problems pertaining to quality that are expressed by the different stakeholders in the construction industry. Products and services will be delivered with highest quality on a timely basis at a reasonable cost to the owner. By implementing the ISO 9000 quality management standards in the construction company measureable results will be achieved with incremental and continuous improvement. The improvement will be visible in functioning of every department in the company. A greater harmony will exist in execution of a construction project leading to higher levels of quality and customer satisfaction with project delivered on time and at a fair price.

Chapter 1 deals with historical review of development of built environment and quality in India. Development of built environment in India is as old as the history of the civilization. The earliest remains of building activity in India dates back to the Indus valley civilization. Among India's ancient architectural remains, the most characteristic ones are temples, chaityas, stupas and other religious structures in ancient India. Temple architecture of high quality workmanship was developed in almost all regions. The distinct architecture style in different parts of the country was a result of geographic, climatic, ethnic, racial and historic diversities. This has led to creation of an epic built environment in the Indian scenario.

From beginning of twentieth century, developmental activities were taken up extensively by Government agencies and included buildings and roads, dams and irrigation projects, water supply and sanitation, housing, railways, airports, etc. Modern structures constructed in pre independence era include Rashtrapati Bhavan, North and South Block in New Delhi and many other buildings, of excellent workmanship, durability and quality. Indian architecture progressed with time and assimilated many influences that came in as a result of India's global discourse with other regions of the world. After Independence of India in 1947, construction works were taken up both through, public sector and private sector agencies which resulted in continuous growth in construction sector over a period of time.

There had been many stages of development in strategies for managing quality in construction works. Inspection for quality was considered appropriate for the quality management in the construction industry. Further quality was also controlled by inspections, monitoring and feedback. Therefore, quality assurance methods became prevalent in construction works for the implementation of planned and systematic activities to achieve quality. Quality assurance was essentially treated as a preventive activity against defects and had to be systematically planned in advance. This included identification and planning of the checks, inspections and quality assurance measures.

Chapter 2 deals with the historical background of quality. It is important to visualize growth and development of quality management systems in construction industry. Quality used to be synonymous to craftsmen's skill during pre-industrialization period. High quality product was considered a pride of workmanship. The advent of industrial revolution and evolution of different concepts of quality laid the foundation of modern quality management systems. The industrial revolution led to the transition from craftsman made limited number of products to the mass manufacturing through machines. The second industrial revolution started in 1870, primarily in Britain, Germany and United States, as also in France, Italy and Japan. It later spread throughout Western Europe and continued till the start of World War I in 1914. Quality systems started developing since beginning of 20th century and became a critical element of business in Japan after World War II, and later a vital business element in US during 1980 onwards.

Many prominent Quality Gurus such as Shewhart, Deming, Juran and others have emerged within the quality field, but some have stood out as key figures for developing different aspects of quality. Their ideas, concepts, and methods are brought out in brief to explain evolution of thinking about quality over a period of time. Later on, International Organization for Standardization (ISO) played an important role in industry removing barriers in trade between nations by publishing international standards. The ISO 9000 family of standards on quality were first published in 1987. Presently, quality is underlying concept to improve performance of business organizations and is required to be built into the products and services. The construction industry is lagging behind in their realization that the customer satisfaction is key to success, and that through planned activities, an organization can be made more efficient, productive and profitable while producing a superior quality at optimum price.

Chapter 3 deals with ISO 9000 Family and Related Standards. It is worth mentioning that prior to introduction of ISO 9000 standard, ISO 8402 Quality Management and Quality Assurance Vocabulary standard was used. Current edition of standards are: ISO 9000:2015-QMS—Fundamentals and Vocabulary, ISO 9001:2015-QMS—Requirements and ISO 9004:2018—Quality Management–Quality of an Organization–Guidance to achieve sustained success. ISO 9000 standard deals with definitions and terminology used to explain concepts used by ISO 9001 and ISO 9004 standards. The ISO 9001:2015-QMS is a business management tool which creates a way of doing business in present day context. The ISO 9001:2015 management system is applicable to each and every aspect of the organization, i.e. from understanding client's requirement to delivery and post-delivery activities. The organization's efficiency, output and quality are aligned and executed optimally by use of ISO 9001 model. It provides a structure through which the management of an organization can control the resources and output of its business in an optimum manner as also monitor quality. The ISO9001 business model outlines the organization's vision, mission, strategies, infrastructure, organizational structure and operational procedures. It supports the delivery of products and services through application of effective and continually improving systems, whilst enhancing customer satisfaction.

ISO's Annex SL Directives represents a framework which provides guidance as to how future ISO Management System Standards (MSS) should be written. The aim of Annex SL was to enhance the consistency and alignment of MSS by providing a unifying and specified high level structure, identical core text and common terms and core definitions. These directives state that all management system standards will use a consistent structure as also common text and terminology. On account of Annex SL directives, ISO 9001:2015 became consistent in structure with other Management System Standards (MSS)—such as ISO14001:2015—Environment Management System and, ISO 45001:2018—Occupational Health and Safety Management System Standard. Other important relevant standard for construction industry includes, ISO 19011:2018—Auditing Management Standard.

Chapter 4 deals with ISO 9000 Quality Management System Requirements and Guidelines, clause wise requirements and their relevance to quality in construction. Benefits to an organization for implementing a Quality Management System based on ISO 9001:2015 are: (i) Product/services meeting customer and regulatory requirements, (ii) enhanced customer satisfaction, (iii) addresses risks and opportunities associated with context and objectives, (iv) demonstrates conformity to specified QMS requirements. In this chapter, consideration has been given to each specified requirement of the ISO 9001 standard and its application to the construction organizations. The chapter covers the guidelines and clarifications for requirements given in ISO 9001 standard.

It is necessary to understand, corresponding provisions of ISO 9000:2015—Quality Management System–Fundamentals and Vocabulary, and ISO 9004:2018-QMS for Sustained Success of Organization. It is important to fully appreciate, specific requirements of ISO 9001:2015. Therefore, in this chapter forclauses of ISO 9001:2015, related and corresponding provisions of these ISO 9000 family standards have been explained.

Chapter 5 deals with implementation agencies of ISO 9001. Accreditation is the formal recognition by an authoritative body regarding the competence to work to specified standards. The apex role for ISO 9001 accreditation authorization is done by International Accreditation Forum (IAF) through Multilateral Recognitions Arrangements (MLA) for mutual recognition across the world. In India, Quality Council of India is performing its pivotal role in the specific areas of accreditation as well as for Quality Promotion. As part of QCI, National Accreditation Board for Certification bodies (NABCB) is signatory of IAF MLA for ISO 9001. Other certification bodies also work in India which are accredited by other equivalent accreditation bodies working in other countries. Bureau of Indian Standards (BIS) is national standard body accredited by NABCB for ISO 9001.

Chapter 6 deals with quality management system auditing. ISO 9000 standards emphasize on the importance of QMS audit and review of its results by management. The terms and definitions related to QMS audit are given in ISO 9000:2015. Audit is also a concept class pertaining to terms and definitions mentioned in ISO 9000

Standard. ISO 19011 standard provides guidelines for auditing management systems which cover guidance on management of audit programme, on planning and conducting of management system audits, as well as on the competence and evaluation of auditor and an audit team. Auditing is characterized by reliance on a number of principles. These principles should help to make the audit an effective and reliable tool in support of management policies and controls, by providing information on which an organization can act in order to improve its performance. The seven principles and processes of auditing covered in ISO 19011:2018 are the same for auditing of different management systems developed on Annex SL framework basis.

The use of "process approach" is a requirement for all ISO management system standards in accordance with Annex SL. Auditors should understand that auditing a management system is auditing an organization's processes and their interactions in relation to one or more management system standard(s). The audit is completed when all planned audit activities have been carried out.

Chapter 7 deals with construction project management practices and quality aspects. Main stages of construction project life cycle consist of project formulation and appraisal, project development, planning for construction, tender action, construction, commissioning and handing over. A construction project is an endeavour of the project team on behalf of owner/client to create a built facility as per defined functional objectives. Quality management in construction aims to achieve required functional and physical characteristics of a constructed facility through meticulous planning and effective quality management practices. Quality management concepts also have a positive effect on time and cost of the project. The vital role of quality management is to ensure that a construction work is able to achieve its full life span with least maintenance costs.

Chapter 8 deals with implementation of ISO 9000 in construction: Case studies of organizations, who have implemented relevant ISO management system standards in their working. Amongst major principal employers, CPWD followed ISO 9000 for their Parliament Library Building Project. Similarly PWD Delhi also followed ISO 9000 standards for their flyover projects. Amongst the fore-running organizations worth mentioning are M/s ICT Pvt. Ltd., New Delhi, a consultancy organization of repute, and M/s Ahluwalia Contracts (India) Ltd., a leading construction organization of the country.

As the customers push for higher levels of quality continues, the efficient construction related companies will became more quality focused and cost competitive. Reorganization of organizational structure and positive attitudes about quality are imperative to an organization in today's scenario. The organization that delivers quality products consistently will continue to grow and prosper. This growth should be enhanced through sound quality management system and its implementation to achieve higher levels of performance and customer's satisfaction.

It is hoped that this book will help in better understanding of ISO 9000 practices in construction. This should help India's construction standard to become equitable with international level of quality standards.

1

Historical Review of Development of Built Environment and Quality in India

OVERVIEW

Development of built environment in India is as old as the history of the civilization. Creating an epic built environment in Indian scenario had a tradition of civil construction of more than 2000 years old. Hindu art traditions included Buddhist and Jain art primarily of temple architecture, chaityas, stupas and other religious structures in ancient India. Islamic art traditions developed during 11th century included buildings like Taj Mahal at Agra, and Gol Gumbaz at Bijapur. Rajput and European art and architecture traditions started developing during 8th century onwards and included primarily forts and palaces. Indian architecture also progressed with time and assimilated the many influences that came in as a result of India's global discourse with other regions of the world. The distinct architecture style in different parts of the country was a result of geographic, climatic, ethnic, racial and historic diversities.

The earliest remains of building activity in India date back to the Indus valley civilization. During the 18th and 19th centuries in the country, developmental activities were taken up extensively by government agencies. Construction works were taken up both through government, public sector and private sector agencies which resulted in continuous growth in construction sector over a period of time.

Quality in initial constructions was considered as intrinsic-skill of master-craftsmen/head workmen. In those days, best construction material and highly skilled workmen were used by rulers in construction and the responsibility of quality was predominantly on skilled workmen only. Later, when civil construction got carried out through contractors by principal employer, the quality concepts developed to another level of quality control and quality assurance. Quality control was focused on detection and remedial action on quality issues whereas quality assurance focus was on preventive action against defects. Even now, quality management has been in vogue in many places in the construction and has evolved to its present from over centuries of construction activity in India.

1.1 HISTORICAL REVIEW: ANCIENT PAST

1.1.1. Indus Valley Civilization

The Indus Valley Civilization dates back to 3300 BC onwards and covered a large area around the Indus River basin and beyond and included development of

Harappa, Lothal, and Mohenjo-daro. The civic, town planning and engineering aspects of these developments were remarkable. The design of these buildings was of utilitarian character, with drains, water courses, tanks and wells. In most sites fired mud bricks were extensively used. Most houses were of two storeys and were of uniform sizes.

1.1.2 Religious Architecture

Since the period of 600 BC–300 BC, walled cities with large gates and low rise multi-storied buildings consistently used arched windows and doors and made an intense use of wood as important feature of architecture. The cities such as Kushinagar and Rajagriha were splendid walled cities of that time.

Religious buildings in the form of the Buddhist stupa, a dome-shaped monument, were used in India as commemorative monuments associated with storing sacred relics of the Buddha. The relics of the Buddha were spread between eight stupas, in Rajagriha, Vaishali, Kapilavastu, Allakappa, Ramagrama, Pava, Kushinagar, and Vethapida.

1.1.3 Classical Period

The next group of buildings, relying on the first examples of true stone architecture, started during the classical period of 300 BC to 6th century AD, with the rise of the Mauryan Empire. The capital city of Pataliputra was an urban marvel described by the Greek ambassador Megasthenes. Remains of monumental stone architecture with a strong Greek influence can be seen through numerous artifacts recovered from Pataliputra.

Fortified cities with stupas, viharas, and temples were constructed during the Maurya Empire. Architectural creations of the Mauryan period, such as the city of Pataliputra and the Pillars of Ashoka, were outstanding constructions in themselves and often compared favourably with the architecture of rest of the world at that time. Stupas were richly decorated with sculptural reliefs.

The Indian emperor Ashoka established the Pillars of Ashoka throughout his realm, generally close to Buddhist stupas. Ashoka also built the Mahabodhi temple in Bodh Gaya around the Bodhi tree under which the Buddha had found enlightenment. Buddha also established slabs of sand stone decorated with reiefs at the foot of the Bodhi tree.

1.1.4 Middle Ages

Indian temple architecture in South India was visible as a distinct tradition of Māru-Gurjara temple architecture which originated somewhere in the 6th century in and around areas south of Rajasthan. This architecture showed the deep understanding of structures and refined skills of the craftsmen of the bygone era.

North Indian temples showed increased elevation of the wall and elaborate spire by the 10th century. Richly decorated temples—including Khajuraho—were constructed in Central India. Grand constructions beautiful sculptures, delicate carvings, high domes, goperas and extensive courtyards were the features of the then temple architecture in India.

The earliest structures of Indo-Islamic architecture were constructed by the Delhi Sultans, most famously the Qutab Minar complex. The complex consists of Qutab Minar, a brick minaret as well as other monuments built by successive Delhi Sultans.

Hoysala architecture was the distinctive building style developed under the rule of the Hoysala Empire in the region of Karnataka, between the 11th and the 14th centuries. Large and small temples built during this era remained as examples of the Hoysala architectural style, including the Chennakesava Temple at Belur, the Hoysaleswara Temple at Halebid, and the Kesava Temple at Somanathapura. Other examples of fine Hoysala craftsmanship were the temples at Belavadi, Amrithapura, and Nuggehalli.

Vijayassacer architecture during 14th and 15th centuries AD was evolved by Vijayanagar empire that ruled from their capital at Vijaya Nagar on the banks of Tungbhadrariver. The Vijayanagar architecture style had a combination of the Chalukya, Hoysala, Pandya and Chola styles which evolved in the earlier centuries.

The most famous Indo-Islamic style was Mughal architecture. Its most prominent examples were the series of imperial mausolea, which started with the pivotal Tomb of Humayun. The Red Fort at Agra and the walled city of Fatehpur Sikri were among the top architectural achievements of that time—as was the Taj Mahal, built as a tomb for Queen Mumtaz Mahal by Shah Jahan. The architecture during the Mughal Period, with its rulers being of Turco-Mongol origin, had shown a notable blend of Indian and Islamic styles.

1.1.5 British Era

The British arrived in 1615 and were present in India for over three hundred years. Their legacy remained through some buildings and infrastructure that were in the country. The major cities colonized during this period were Madras, Calcutta, Bombay and Delhi.

Mumbai (then known as Bombay), had some of the most prominent examples of British colonial architecture. This included Victoria Terminus, University of Mumbai, Rajabai Clock Tower, High Court, BMC Building of Wales Museum, Gateway of India and Taj Mahal Palace Hotel. The Victoria Memorial in Calcutta was one of the most effective symbolism of British Empire. Parliament House and Central Secretariat were also constructed in New Delhi in this era.

1.2 POST-INDEPENDENCE SCENARIO

After Independence, the important buildings constructed included Supreme Court of India at Delhi and Vidhan Sabha at Bangalore. In 1950, French architect Le Corbusier, a pioneer of modernist architecture, was commissioned to design the city of Chandigarh. The city plan included residential, commercial and industrial areas, along with parks and the transportation infrastructure. There was significant increase in developmental activities post-Independence, mainly by the government agencies as a part of planned economic development.

The construction works increased many fold in all fields including, roads, buildings, irrigation, dams, water supply, sanitation, railways, airfields, etc. These were launched by different agencies of central and state governments. Important milestones of this planned economic development can be seen in many mega projects, and specific reference could be made of Bhakra Dam near Chandigarh.

1.3 QUALITY IN CONSTRUCTION

1.3.1 General

In the ancient past, major construction works were carried out through the rulers of that period, using best construction materials and by employing best skilled labour available in the country. In such periods, quality of construction was never an issue as the master craftsmen were responsible for inspecting and monitoring quality of their own work. Master craftsmen were employed and their trade got transferred from generation to generation, retaining their intrinsic skill.

Subsequently the civil constructions were got carried out through contractors with principal role for quality of work being carried through the agency executing the work. They were expected to arrange for required materials and labour and produce the desired quality in construction. The quality of construction work was controlled through inspection of materials, workmanship and adherence to the process of construction methodology for various items of work. The agency was required to engage staff with required knowledge, experience and competence.

The engineers/architects/inspectors supervising a construction project need to be familiar with drawings, designs and provisions of the contract. This includes familiarity with plans, specifications, method of construction and all revisions and amendments to the contract. With increased complexity of the projects and with the stakes of the employer getting higher in timely completion of work and soundness of quality, it becomes necessary for management to have an active role in construction process. Quality-oriented inspections are therefore necessary to ensure that the construction work achieves the desired quality. Quality is generally considered as conformance to specifications and is based on detection of defects at the execution stage of the project including remedy of the said defects.

When the construction works are carried out by the Government agencies like CPWD, MES, Indian Railways, etc. the role of the engineers executing the construction work is critical in respect of quality of materials, quality of workmanship and overall requirements of quality. The engineers have to critically review, evaluate and analyse, and take appropriate measures to ensure quality of construction. Irrespective of the agency of construction, whether done departmentally or through a contractor, inspection of works and testing of material and workmanship is the principal mode of maintaining quality of work at site. The engineer needs to look upon and monitor critically the execution of the project and take appropriate action for management of quality.

Quality control (QC) is an important function of quality in construction industry. It encompasses all the techniques and activities of the organization that continually monitor the construction quality. QC aims to reduce the variation of product quality. QC aims to eliminate the causes of unsatisfactory product performance, and to review and correct the sources of variation.

Quality assurance requires planned and systematic actions necessary to assure that construction will meet the specific quality requirements. Quality assurance ensures that all activities from planning, design, development and execution of work have been performed to achieve customer satisfaction. The activity also includes identification and planning of checks, inspections and control process, including establishment of a quality management system which can help in "doing things right the first time".

A contractor's in house quality management system is also of utmost importance. An internal quality system provides confidence to the management of an organization that the intended quality is being achieved. Successful implementation of quality management systems can contribute to an increase in product quality, improvement in workmanship and efficiency, a decrease in wastage and increased profits.

1.3.2 Role of Quality Management in Future Construction Projects

Presently the construction companies have to deal with several complex tasks to grow successfully, to stay competitive in a globalised economic environment and to satisfy their customers' requirements and expectations. The knowledge of quality management has developed worldwide which includes many changes from the past. It is the need of the hour that construction industry accepts the principles and concepts of present quality management systems and enhance quality in construction to remain at par with the quality in other developed nations.

At this point it is appropriate to mention that the authors are focusing on the historical aspects of construction as well as evolution of latest quality concepts in construction and project management, consistent with ISO Quality Management System Standards. This offers step by step guidance on implementation and management of quality system and demonstrate how the system puts the quality management processes into effect before work begins thereby ensuring required quality in construction projects.

BIBLIOGRAPHY

1. Buddist Architecture, Lee Huu Phucc, Graficol 2029.

2. Encyclopedia Britannica (2009), toril

3. Gast, Klaus_Peter (2007), Modern Tradition: Contemporary Architecture in India, Birkhauser, ISBN 978-3-7643-775-0.

4. Harley JC. The ART and Architecture of the India Subcontinent, 2nd end., 1994., Yale University Press, Pelican Hisotry of Art, ISBN 0300062176.

5. https://en.wikipedia.org/wiki/architecture of India

6. Mitchell, George, The Hindu Temple Architecture: An Introduction to its meaning and forms, 1977, University of Chicago Press, ISBN 978-0-226-53230-1.

7. Thapar, Bindia (2004): Introduction to Indian Architecture, singapore: Periplus Editions, ISBN 978-0-7946-0011-2.

2

Historical Development of Quality Management

OVERVIEW

Quality in products and services was synonymous to craftsmen's skill during pre-industrial revolution period. During that period every effort was made to build quality in the product by people who were involved in production. High quality in product was considered as pride of workmanship. The advent of industrial revolution and evolution of different concepts of quality laid the foundation of modern quality management. The industrial revolution led to the transition from craftsman made limited number of products to the mass manufacturing through machines. The commencement of the industrial revolution is closely linked to a number of innovations during the second half of the 18th century. Second industrial revolution started in 1870, primarily in Britain, Germany and United States, as also in France, Italy and Japan. The second industrial revolution later spread throughout Western Europe and continued till the start of World War I in 1914. Quality systems started developing since beginning of 20th century and became a critical element of business in Japan after World War II, and later a vital business element in US during 1980.

Many prominent quality gurus such as Shewhart, Deming, Juran and others have emerged within the quality field, but some have stood out as key figures for developing different aspects of quality. Their ideas, concepts, and methods are brought out in brief to explain evolution of thinking about quality over a period of time. Later on, International Organization for Standardization (ISO) played an important role in industry removing barriers in trade between nations by publishing international standards. The ISO 9000 family of standards on quality were first published in 1987.

Presently, quality is underlying concept to improve performance of business organizations and is required to be built into the products and services. The construction industry is lagging behind in their realization that the customer satisfaction is key to success, and that through planned activities, an organization can be more efficient, productive and profitable while producing a superior quality at optimum price.

2.1 QUALITY IN EARLY 20TH CENTURY

2.1.1 In the year 1911, Frederick W Taylor introduced "Scientific Management Principles" which were predominantly focused on enhancement of productivity in manufacturing units. His philosophy led to the responsibility of planning to managers and engineers, execution to workers and quality to the quality inspectors, who inspect the manufactured product. Consequently, manufacturing companies started creating a separate quality inspection department to check the conformance of standards in products. The concept of final product inspection at the end of manufacturing process was the primary tool of quality control up to the first half of 20th century. The key aspect of quality inspection was to detect the quality problems in manufactured products and correct it afterwards.

2.1.2 Walter A. Shewhart spent his professional carrier at Western Electric and Bell Telephone Laboratories, both divisions of AT and T. In 1924, he developed the control chart theory which was focused on application of statistical methods for quality control. The control chart employed statistical methods to monitor and control a process and it emphasized early detection and prevention of defects, rather than the correction of the same after they occurred. Graphic representation of control chart is shown in Fig. 2.1. Defects due to normal variation remain within upper and lower control limit. Defects on account of special causes are out of control limit. Shewhart's work pointed out the importance of reducing variation in a manufacturing process and the understanding that continual process-adjustment in reaction to non-conformance actually increased variation in product.

Shewhart concluded that while every process displays variation, some processes display controlled variation that is natural to the process, while others display uncontrolled variation that is not present in the process control system at all times. Control charts attempt to differentiate assignable special sources of variation from common cause. "Common" causes, are an expected part of the processes, and are of much less concern to the manufacturer than "assignable" sources. Eliminating causes of variation are significant for process stability. In

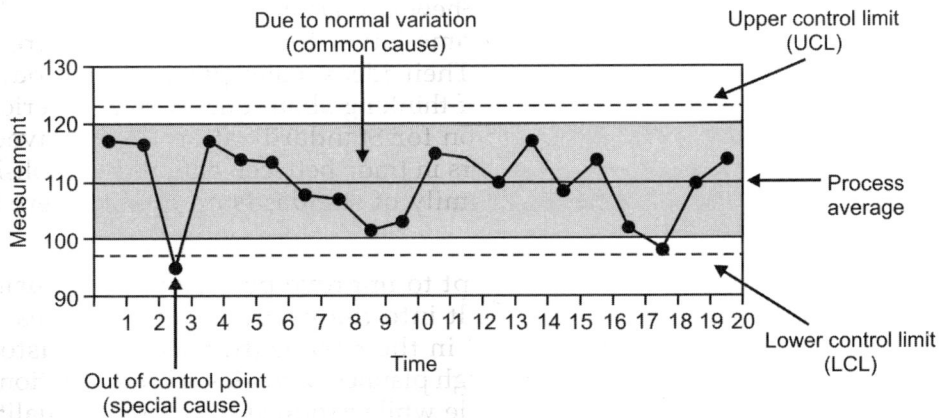

Fig. 2.1: Control chart

1935, Shewhart published a book on statistical quality control, titled "Economic Control of Quality of Manufactured Products". This book was regarded as a complete and thorough work on the basic principles of quality control.

2.1.3 In 1930, Harold Dodge and Harry Romig statisticians at Bell Labs, developed acceptance sampling tables to determine whether to accept or reject a production lot of material. The sampling plan of Dodge-Romig was optimum inspection plan and worked as alternative to 100% inspection of products. The lot is accepted if the number of defects fall lower than the specified acceptance number or otherwise the lot is rejected. It became a common quality control technique used successfully in the industry.

2.1.4 In 1939 Second World War started, when quality became a critical component of the war effort and an important safety issue. The armed forces began to use sampling inspection instead of unit-by-unit inspection. The acceptance sampling tables were published by them in the military standard, known as MIL-STD-105. These tables were incorporated into the military contracts so that suppliers could understand what they were expected to produce. The armed forces also helped suppliers to improve quality, by sponsoring training courses in Walter Shewhart's Statistical Quality Control (SQC) techniques and imposed stringent conditions to their supplier to follow statistical sampling.

2.2 QUALITY MANAGEMENT SCENERIO—POST WORLD WAR II

2.2.1 After the end of Second World War, Japan began to transform its industrial sector through quality movement. From 1950 onwards, Deming, an American quality expert, played an important role in improving quality of products in Japan. He focused on Statistical Process Control training and application. Deming made a significant contribution to establish Japan's reputation for innovative, high-quality products, and in turn it emerged as an economic power. He said, "Good quality does not necessarily mean high quality. It means a predictable degree of uniformity and dependability, at a low cost, with quality suited to the market." Deming also got popularity for his contribution to improvement cycle or Plan-Do-Check-Act (PDCA) cycle. PDCA is a four-step management method used for the control and continuous improvement of processes and products, as detailed below:

Plan   Establish the objectives and processes necessary to deliver results in accordance with customer requirements and the organization's policies.

Do   Implement the processes

Check   Monitor and measure processes and products against policies, objectives and requirements for the product and report the results

Act   Take action to continually improve process performance

2.2.2 From 1954 onwards, another American quality expert Dr Joseph M. Juran played pivotal role by motivating as well as educating Japanese higher level management for quality improvement initiatives. He focused on managing for quality and is widely credited for adding the human dimension to quality management. He started the education and training of managers in the field of quality. Dr Juran raised the level of quality management from the factory to that of the total organization. He stressed the importance of systems with product design, prototype testing, proper equipment operations and accurate process feedback. He also defined quality as "Fitness for use". Dr Juran provided the change over from Statistical Quality Control (SQC) to Total Quality Control (TQC) in Japan. This included company-wide TQC activities and related education in quality control. This quality driven approach helped Japanese products to significantly penetrate into Western market. Dr Juran also mentioned about the cost of quality. He defined quality of a product by its features and absence of deficiencies. This was illustrated by his "Juran trilogy," an approach to cross-functional management, which is composed of three managerial processes, i.e. quality planning, quality control, and quality improvement. His philosophy was that without change, there will be a constant waste, during change there will be increased costs but after the improvement, margins will be higher and the increased costs shall get recovered.

In 1992, Dr Juran's book *Quality by Design* was published which describes what is required to achieve breakthrough in new products, services, and processes. He defined quality by product features and absence of deficiencies. The steps for quality by design model are: (a) Establish the project design targets and goals, (b) define the market and customers that will be targeted, (c) discover the market, customers, and societal needs, (d) develop the features of the new design that will meet the needs, (e) develop or redevelop the processes to produce the features and (f) develop process controls to be able to transfer the new designs to operations.

2.2.3 Kaoru Ishikawa is one of the world's foremost authorities on quality control. He said that through total quality control with the participation of all employees, including the president, any company can create better product (or service) at a lower cost, increase sales, improve profits and turn the company into a better organization. He propagated "to practice quality control to develop, design, produce and service a quality product which is economical, useful and meets consumer satisfaction. To meet this goal, everyone in the company must participate including top executives, all divisions, and all employees." One of the innovative quality improvement through TQC methodology was developed by Kaoru Ishikawa through cause-and-effect" diagram. In Fig. 2.2, the effect is shown as the fish's head, facing to the right, with the causes extending to the left as fishbone; the ribs branch off the backbone for major causes, with sub-branches for root-causes, to as many levels as required. Ishikawa's diagram lead Japanese firms to focus improvement through quality control of materials, pieces of equipment, and processes. He envisaged that by using cause and effect diagram (also called the "Ishikawa" or "Fishbone" diagram) the management leaders could make significant advancements in quality improvement.

2.2.4 In 1962, Quality Circle concept was also introduced in Japan by Kaoru Ishikawa. A quality circle or quality control circle is a small group of workers, led by supervisor or manager, who do the same or similar work, who meet regularly to identify, analyze and solve work-related problems. Later on, Kaoru Ishikawa outlined the elements of Total Quality Control (TQC) management as,

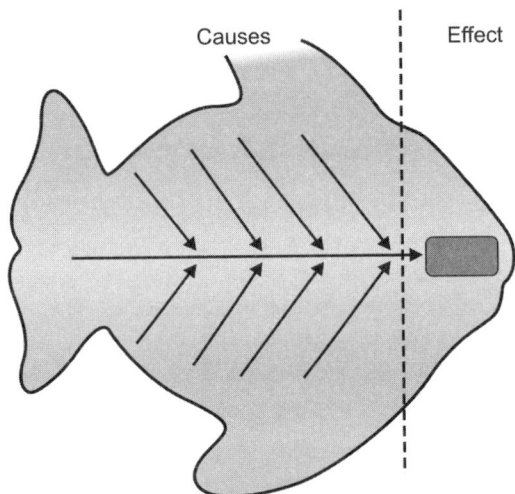

Fig. 2.2: Cause-and-effect diagram/fish bone diagram

(a) quality comes first, not short-term profits, (b) the customer comes first, not the producer, (c) decisions are based on facts and data, (d) management is participatory and respectful of all employees, (e) management is driven by cross-functional committees covering product planning, product design, production planning, purchasing, manufacturing, sales, and distribution.

2.2.5 The evolving concepts of total quality control and zero defect, progressively led to the phase of quality assurance which was predominantly dependent on two principles, the first principle was given by Dr Joseph M Jurans "Fit for purpose" (the product should be suitable for the intended purpose); and another one was given by Philips Crosby—"right at the first time", it means mistakes should be eliminated. Quality assurance includes management of the quality of raw materials, assemblies, products and components, services related to production, production and inspection processes. Quality assurance (QA) is a way of preventing mistakes and defects in manufactured products and avoiding problems when delivering products or services to customer. This defect prevention in quality assurance differs from defect detection and rejection in quality control. Quality assurance comprises administrative and procedural activities implemented in a quality system so that requirements and goals for a product, service or activity will be achieved.

2.2.6 In 1964, Philips Crosby added a new dimension to quality assurance, known as "Zero Defect" (ZD). He defined quality as conformance to requirements which include both customer's requirement and product's requirement. It was a management-led program to eliminate defects in industrial production through prevention. It is directed at motivating people to prevent mistakes by developing a constant, conscious desire to do their job right the first time. Afterwards, Crosby also wrote a book *Quality is Free*. Crosby divided quality-related costs into the price of conformance and the price of nonconformance. The price of conformance includes quality-related planning, inspection, and auditing; the price of non-conformance includes scrap, rework, claims against warranty and unplanned service. "Quality is Free" demonstrated that quality improvement efforts pay for themselves.

2.2.7 During the late 1970s and early 1980s, the developed countries of North America and Western Europe suffered economically in the face of stiff competition from Japan's ability to produce high-quality goods at competitive cost. Japanese were seen as the leaders in quality. For the first time since the start of the industrial revolution, the United Kingdom became a net importer of finished goods. Western world began re-examining the techniques of quality control developed over the past 50 years and how those techniques were successfully employed by the Japanese. Thus rethinking emerged in the United States regarding the quality issue.

2.3 POST-NINETEEN EIGHTY—US "QUALITY REVOLUTION"

2.3.1 From 1980 onward, the competition in business grew rapidly due to various factors such as technological advancement as also increased awareness of quality in consumers, industry and government. Business groups began to understand that quality is the key driver for survival in worldwide competitiveness. This was the time when worldwide interest in quality grew immensely. During this period quality revolution started in America and consumer focused quality took precedence over product focused quality.

2.3.2 American television episode broadcasted "If Japan can Why can't we" by NBC News as part of the television show "NBC White Paper" on June 24, 1980. This show was credited as the beginning of Quality Revolution in America. W. Edwards Deming got popularity in America after this TV program. Deming said that 85% of a worker's effectiveness is determined by the system he works with, only 15% by his own skill.

2.3.3 US industry and government leaders realized that a renewed emphasis on quality was necessary for doing business in an expanding and competitive world market. The CEOs of major US corporations came forward to provide personal leadership in the quality movement. The US initiative, emphasizing not only statistics but approaches that embraced the entire organization, this was known as Total Quality Management (TQM). TQM is an organization-wide effort to "install and create suitable environment where employees continuously improve their ability to provide on demand products and services that customer will find of significant value. "Total" emphasizes that all functional departments in an organization are obligated to improve their operations.

2.3.4 In 1980, Noriaki Kano developed a model for customer satisfaction (now known as the Kano model) which was quite helpful in product development. Kano believed that not all attributes of product or services are equal in the eyes of the customer, and some specific attributes create higher levels of customer loyalty than others. Kano model focuses on differentiating product features, as opposed to focusing initially on customer needs. The model illustrates three levels of customer satisfaction, i.e. threshold, performance and excitement for different product characteristics, as depicted in Fig. 2.3.

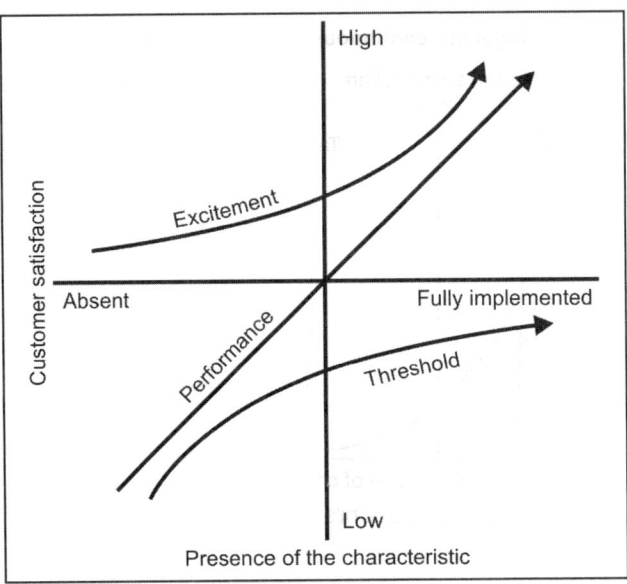

Fig. 2.3: Kano model

2.3.5 Genichi Taguchi defined quality as the ill effects (loss to the society) of lower-quality products in the hands of consumers as depicted in Fig. 2.4. He surpassed the common thinking around specification limits that the customer is satisfied as long as the variation stays within the specification limits. The specification limits divide satisfaction from dissatisfaction. Taguchi states that any variation away from the nominal (target) performance will start customer dissatisfaction. As the variation increases, the customer will gradually (exponentially) become dissatisfied. The loss function is also depicted graphically in Fig. 2.5. Taguchi further emphasized on robust design.

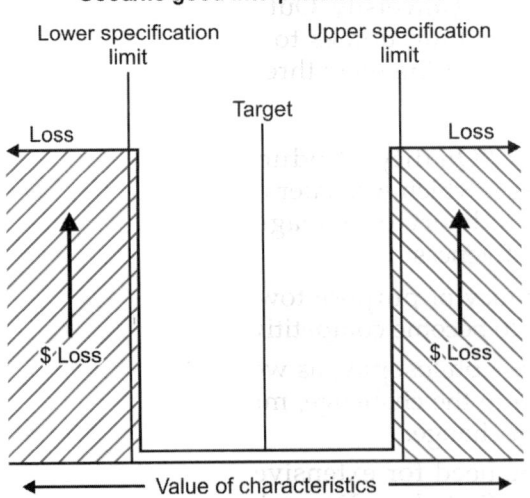

Fig. 2.4: General interpretation of loss

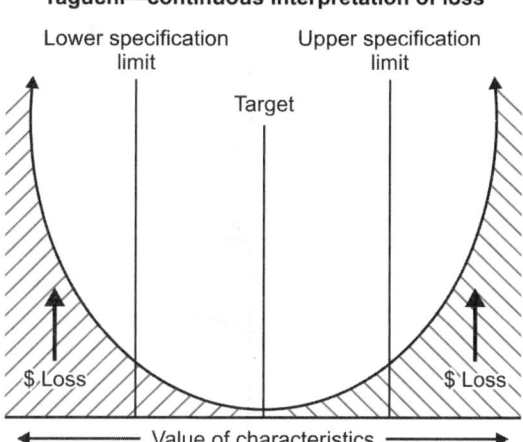

Fig. 2.5: Taguchi's interpretation of loss

2.3.6 Bill Smith an engineer working at Motorola introduced a set of techniques and tools for improvement in process termed as "Six Sigma". Six Sigma is a disciplined, statistical-based, data-driven approach and continuous improvement methodology for eliminating defects in a product, process or service. Six sigma also means less than 3.4 defects to a million operations. If a defect is defined by specification limits separating good from bad outcomes of a process, then a six sigma process has a process mean (average) that is six standard deviations from the nearest specification limit. This provides enough buffer between the process natural variation and the specification limits. Jack Welch adopted Motorola's Six Sigma quality at General Electric (GE) in late 1995. One of the example of Six Sigma in India is Mumbai Dabbawall as. They deliver food every day from the home to the workplace and they do it with minimal technology, processes, or structure. They became world renowned as a Six Sigma Organization, and were studied by Harvard University. Out of every million lunch boxes that need to be reached from a million homes to their respective destinations in different offices in Mumbai, at the most three could miss reaching the right desk at the right time.

2.3.7 Deming's book "Quality, Productivity, and Competitive Position", was published in 1986, which was later on renamed it as "Out of the Crisis". In this book he offers a theory of management based on 14 points to improve the company's effectiveness:

a. Create constancy of purpose toward improvement of product and service, with the aim to become competitive, to stay in business and to provide jobs.

b. Adopt the new philosophy, as we are in a new economic age. Management must awaken to the challenge, must learn their responsibilities, and take on leadership for change.

c. Eliminate the need for extensive inspection, by building quality into the product in the first place. Dependence on final product inspection to achieve quality to be avoided.

d. End the practice of awarding business on the basis of a price tag. Instead, minimize total cost. Move towards a single supplier for any one item, on a long-term relationship of loyalty and trust.

e. Improve continuously the system of production and service, to improve quality and productivity, and thus constantly decrease costs.

f. Institute training on the job.

g. Institute leadership, the aim of supervision should be to help people and machines and gadgets to do a better job. Supervision of management is in need of overhaul, supervision of production workers is also in need of overhaul.

h. Drive out fear, so that everyone may work effectively for the company.

i. Breakdown barriers between departments. People in research, design, sales, and production must work as a team, to foresee problems of production and usage that may be encountered with the product or service.

j. Eliminate slogans and targets for the work force asking for zero defects and new levels of productivity. Such slogans only create adverse relationships, as the bulk of the causes of low quality and low productivity belong to the system and thus lie beyond the power of the work force.

k. Eliminate work standards on the factory floor and institute leadership.

l. Remove barriers to pride of workmanship. Let people be proud of their work.

m. Institute a vigorous program of education and self-improvement.

n. Put everybody in the company to work to accomplish the transformation. The transformation is everybody's job.

2.3.8 Deming is also known in US for his "System of Profound Knowledge". He said that every company deals with universal knowledge, i.e. profound knowledge, which penetrates a company from the outside. The system of profound knowledge is made up of four components through which the world is looked at simultaneously:

a. **Appreciation of a system:** Understanding the overall processes involving suppliers, producers, and customers (or recipients) of goods and services.

b. **Knowledge of variation:** The range and causes of variation in quality, and use of statistical sampling in measurements.

c. **Theory of knowledge:** The theory of knowledge teaches that a statement, if it contains knowledge, predicts future outcomes, with risk of being wrong. Rational predictions require theory and builds knowledge through systematic revision and extension of theory based on comparison of prediction with observation.

d. **Knowledge of psychology:** Concepts of human nature.

2.4 DEVELOPMENT OF INTERNATIONAL STANDARDS ON QUALITY

2.4.1 The International Organization for Standardization (ISO) was founded on 23rd February 1947 by representatives of 25 countries. India is one of the founder members of ISO and is represented by Bureau of Indian Standards (BIS) in ISO. The member organizations represent the interest of the vendors, manufacturers, consumers, professionals and government. ISO headquarter is in Geneva, Switzerland and works in 164 countries. The ISO promotes worldwide proprietary, industrial and commercial standards. ISO Standards are based on global expert opinion, worldwide consensus, developed through multi-stakeholder process and respond to the need of the market. It is the world's largest developer of voluntary international standards and facilitates world trade by providing common standards between nations. The ISO standards are International Standards for use of business organizations, government and Society as a whole. The Standards help in the development of products and services that are safe, reliable and of required quality. However, ISO itself does not certify business organizations for quality.

2.4.2 To understand the history of first ISO standards of quality we need to look back to the development of quality standards developed in the world. US Military Standard MIL-Q-9858 was published in the year 1959 which later became the basis of BS 5750 standards published by British Standards Institution (BSI) in 1979. The main theory of BS 5750 is that by writing down, and adhering to, set procedures a high quality product or service will be produced. Three quality assurance models of BS 5750 were adopted by ISO Technical Committee (TC-176) and based on that first ISO 9000 family of standards (i.e. ISO 9001, ISO 9002 and ISO 9003) were published on quality assurance during 1987.

2.4.3 Quality System Standards (1987)—ISO 9001, ISO 9002 and ISO 9003

These standards were well-suited to manufacturing and emphasized on conformance to procedures, the selection of which was based on the scope of activities of the organization. These quality assurance model-based standards were published by ISO against which companies could get certified and audited as well. These standards were intended to be used by procuring organizations, as the basis of contractual arrangements with their suppliers. This helped reduce the need for "supplier development" by establishing basic requirements for a supplier to assure product quality. The emphasis was given on conformance to procedures rather than the overall process of management. The purpose of these standards is detailed below:

a. ISO 9001:1987 was for quality assurance in design, development, production, installation, and servicing. The standard was useful for organizations whose activities included the creation of new products including design and development activities.

b. ISO 9002:1987 was for quality assurance in production, installation, and servicing. The standard was similar to ISO 9001 but meant for organizations whose activities did not cover design and development activities.

c. ISO 9003:1987 was for quality assurance in final inspection and test. The standard was meant for companies which cover only the final inspection of finished product, with no concern for how the product was produced.

2.4.4 Quality System Standards (1994)—ISO 9001, ISO 9002, ISO 9003 and ISO 9004-1

a. The ISO 9001:1994, 9002:1994 and 9003:1994 were revised with emphasis on quality assurance via preventive actions, instead of just checking final product, and continued to require evidence of compliance with documented procedures. These standards had generic application. These standards stipulate requirement of procedures, which in turn created shelf-loads of procedure manuals. In some companies, adapting and improving processes could actually be impeded by the quality system.

b. ISO 9001:1994 standard had 20 procedures and specified 20 clauses related to design, development, production, installation and servicing: These are (a) management responsibility, (b) quality system, (c) contract review, (d) design control, (e) document and data control, (f) purchasing, (g) control of customer supplied product, (h) product identification and traceability, (i) process control, (j) inspection and testing, (k) control of inspection, measurement and testing equipment, (l) inspection and test status, (m) control of non-conforming product, (n) corrective and preventive action, (o) handling, storage, packaging, preservation and delivery, (p) control of quality records, (q) internal quality audits, (r) training (s) servicing and (t) statistical techniques

c. ISO 9002:1994 standard specified 18 clauses covering production and installation. These clauses are similar to ISO 9001 except the first two clauses pertaining to design and development. This was applicable for the units other than research and development units.

d. ISO 9003:1994 standard specified 12 clauses covering final inspection and testing for laboratories and warehouses, etc.

e. ISO 9004-1:1994 standards gave guidelines for manufacturing sector. It provided guidance on quality management and quality system elements suitable for use in the development and implementation of a comprehensive and effective in-house quality system. This part (ISO 9004 part-1) was not intended for contractual, regulatory or certification use. References to a 'product' in this standard should be interpreted as applicable to the generic product categories of hardware, software, processed materials or service.

Other Standards in ISO 9000 Series

i. ISO 8402: Quality Management and Quality Assurance Vocabulary

ii. ISO 9000-1: Quality Management and Quality Assurance Standards

    Part 1: Guidelines for Selection and Use

iii. ISO 9000-2: Quality Management and Quality Assurance Standards

    Part 2: Generic Guidelines for Application of ISO 9001, ISO 9002 and ISO 9003

iv. ISO 9000-3: Quality Management and Quality Assurance Standards

    Part 3: Guidelines for Application of ISO 9001 to the Development, Supply and Maintenance of Software

v. ISO 9000-4: Quality Management and Quality Assurance Standards

    Part 4: Guide to Dependability Program Management.

vi. ISO 9004-2: Quality Management and Quality System Elements, part 2, Guidelines for Services

vii. ISO 9004-3: Quality Management and Quality System Elements, part 3, Guidelines for Processed Material

viii. ISO 9004-4: Quality Management and Quality System Elements, part 4, Guidelines for Quality Improvement

2.4.5 Quality Management System Standards (2000)—ISO 9000, ISO 9001 and ISO 9004

a. ISO 9000:2000—Quality Management System—Fundamentals and Vocabulary
ISO 8402 was used for quality vocabulary before publication of ISO 9000:2000 standard. ISO 9000 standard contains terms and definitions related to QMS and also defines eight quality management principles. These principles form the basis for the quality management standards within the ISO 9000 family. The principles along with their statements are given below:

   i. Customer focus: Organizations depend on their customers and, therefore should understand current and future customer needs, should meet customer requirements and strive to exceed customer expectations.

   ii. Leadership: Leaders establish unity of purpose and direction of the organization. They should create and maintain the internal environment in which people can become fully involved in achieving the organization's objective.

   iii. Involvement of people: People at all levels are the essence of an organization and their full involvement enables their abilities to be used for the organization's benefit.

   iv. Process approach: A desired result is achieved more efficiently when activities and related resources are managed as a process.

   v. System approach to management: Identifying, understanding and managing interrelated processes as a system contributes to the organization's effectiveness and efficiency in achieving the objectives.

   vi. Continual improvement: Continual improvement of the organization's overall performance should be a permanent objective of the organization.

   vii. Factual approach to decision making: Effective decisions are based on the analysis of data and information.

   viii. Mutually beneficial supplier relationship: An organization and its suppliers are interdependent and a mutually beneficial relationship enhances the ability of both to create value.

b. ISO 9001:2000—Quality Management System Requirements

   i. Quality management system requirements sought to make a radical change in thinking by evolving the concept of process management. Process approach is the application of system of processes within an organization, together with identification and interaction of these processes, and their management to produce desired outcome. This version also demanded involvement by senior executives in order to integrate quality into the business system and avoid delegation of quality functions to junior functionaries. Expectations of continual process improvement and tracking customer satisfaction were also made explicit.

   ii. This was first major revision of standard which replaced and unified all three former standards of 1994 version, i.e. ISO 9001, ISO 9002 and ISO 9003 standards. There are some significant changes to the requirements.

This is the standard which assessment bodies used for surveillance or certification purposes. The standard structure specified eight clauses: (a) Scope, (b) normative references, (c) terms and definitions, (d) quality management system, (e) management responsibility, (f) resource management, (g) product realization, (h) measurement, analysis and improvement

iii. The standard requires six mandatory procedures for: (a) Control of documents, (b) control of records, (c) internal audit, (d) control of non-conforming product, (e) corrective action, (f) preventive action.

The model of process-based quality management system illustrates process linkages presented in requirements related clauses of ISO 9001:2008. This process model of QMS also shows that customers play a significant role in defining requirements as inputs as well as for feedback information on delivered product. The activities related to product realization have been considered as value added activities. The information flow between management and customer, has been given higher weightage for attaining customer satisfaction. The ultimate objective of this model is to attain continual improvement in product and services by use of clauses, viz. Clause 5: Management Responsibility, Clause 6: Resource Management, Clause 7: Product Realization and Clause 8: Measurement, Analysis and Improvement of the ISO 9001 standard. This model is based on improvement (Plan-Do-Check-Act) cycle which broadly means top management is responsible for "Plan", i.e.

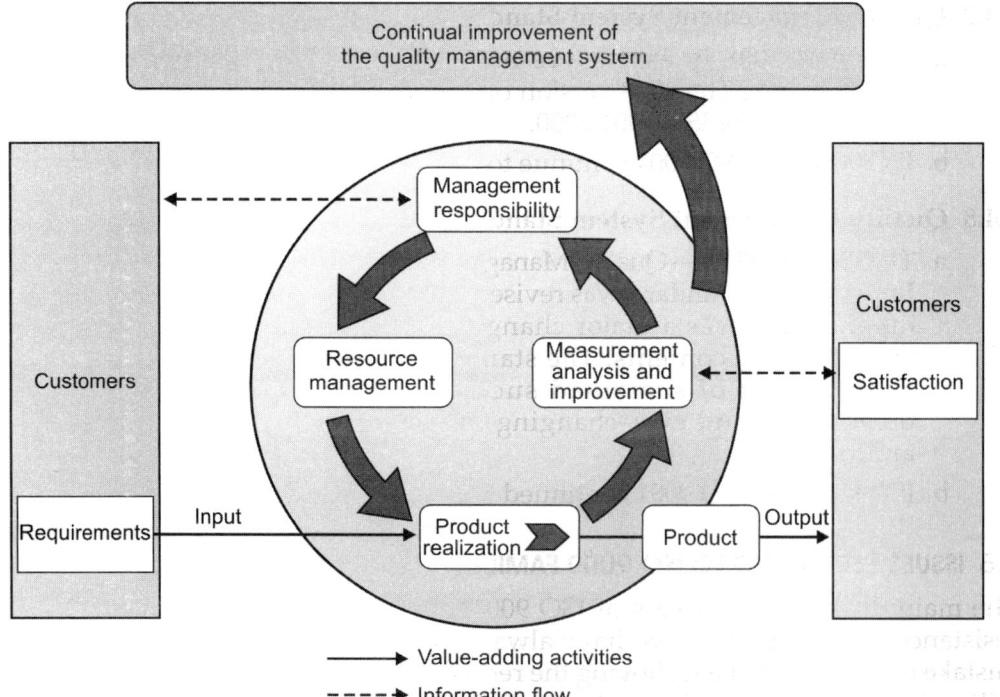

Fig. 2.6: Model of a process-based quality management system

planning of QMS and quality objectives including, resource management and planning related to human resource, training, etc. "Do" here means product realization activities, "Check" and "Act" mean requirements of Clause 8: Measurement, Analysis and Improvement.

c. ISO 9004:2000—Quality Management System—Guidelines for Performance Improvement.

This second edition of ISO 9004 cancels and replaces ISO 9004-1:1994. The title has been modified to reflect the comprehensiveness of the quality management system, standard uses broader perspective of quality management and is more consistent with ISO 9001 compared to previous edition.

2.4.6 Quality Management System Standards (2005)—ISO 9000

a. The ISO 9000:2000 version was further revised in the year 2005 to align the definitions of various terms in ISO 9000 to that of auditing management standard and also with the terms relating to quality assurance for measurement processes. ISO 9000:2005 replaces ISO 9000:2000 standard. ISO 9000:2005 contains 10 concept diagrams of the vocabulary for concepts relating to quality, management, organization, process and product, characteristic, conformity, documentation, examination, audit, quality management for measurement processes. The concept diagrams explain the relationship between related terms in the same concept class.

b. ISO 9001 and ISO 9004 continues to remain same as earlier.

2.4.7 Quality Management System Standards (2008)—ISO 9001

a. ISO 9001:2008 in essence re-narrates ISO 9001:2000 without any new requirements. The 2008 version only introduced clarifications to the existing requirements of ISO 9001:2000.

b. ISO 9000 and ISO 9004 continue to remain same as earlier.

2.4.8 Quality Management System Standards (2009)

a. The ISO 9004:2000—Quality Management System—Guidelines for Performance Improvement standard was revised. Managing for the sustained success of an organization was a major change, leading to substantial changes to its structure and contents. The standard provides guidance to support the achievement of sustained success for any organization in complex, demanding, and ever-changing environment, by a quality management approach.

b. ISO 9000 and ISO 9001 continued to remain same as earlier.

2.5 ISSUES PERTAINING TO ISO 9000 FAMILY OF STANDARDS

The main difficulty in success of ISO 9001 family of standards has been the silent resistance to change—the "we have always done it this way" attitude. The biggest mistake companies make is shoving the requirements down to normal employees and failing in engaging specific employees for the ISO 9001 QMS. Many organizations perceived ISO 9001 QMS just as a quality tool comprising of checklists and procedures and believed the essence of ISO 9001 certification being limited to marketing tool to attract customers. Such organizations generally face the problem of not understanding the ISO 9001 requirements and process approach. Their processes hardly do any value addition to its product or services and managements continue to remain busy in tackling shallow issues and respond to the "problems of the day" rather than addressing process performance issues based on objective measures.

With this approach, they really missed the opportunity to reap the true benefits from ISO 9001 QMS. Consequently, small sized companies continued to remain small and big companies did not achieve optimum profitability. "The good organizations are the ones that make the standard fit the company, not the company fit the standards". It is noteworthy that it is impractical to create a perfect system, which means gaps must be expected and efforts made to reduce gaps with on-going focus on continuous improvement.

BIBLIOGRAPHY

1. Acceptance Sampling- https://en.wikipedia.org/wiki/Acceptance_sampling
2. British Standards -https://en.wikipedia.org/wiki/British_Standards
3. Deming 14 points- https://en.wikipedia.org/wiki/W._Edwards_Deming
4. Deming's principles of TQM- https://people.well.com/user/vamead/demingdist.html
5. History of Japan's quality movement- http://www.wtec.org/loyola/ep/c6s1.htm
6. Improvement cycle - https://en.wikipedia.org/wiki/PDCA
7. Industrial Revolution - https://en.wikipedia.org/wiki/Industrial_Revolution
8. ISO 9000- https://en.wikipedia.org/wiki/ISO_9000
9. Kano model -https://en.wikipedia.org/wiki/Kano_model
10. Malcolm Baldridge National Quality Award - https://en.wikipedia.org/wiki/Malcolm_Baldrige_National_Quality_Award#History_of_the_Baldrige_Program
11. Managing for Quality and Performance Excellence, James R. Evans and William M Lindsay, 9th edition, engage Learning.
12. Noriaki Kano- https://en.wikipedia.org/wiki/Noriaki_Kano
13. Quality Assurance -https://en.wikipedia.org/wiki/Quality_assurance
14. Quality by Design- https://en.wikipedia.org/wiki/Quality_by_Design
15. Quality Circle -https://en.wikipedia.org/wiki/Quality_circle
16. Quality management historyhttps://asq.org/quality-resources/history-of-quality#guilds
17. Second Industrial Revolution-https://en.wikipedia.org/wiki/Second_Industrial_Revolution
18. Six Sigma - https://en.wikipedia.org/wiki/Six_Sigma
19. Statistical Process Control- https://en.wikipedia.org/wiki/Statistical_process_control
20. Statistical Quality Control https://www.itl.nist.gov/div898/handbook/pmc/section1/pmc11.htm
21. Tagauchi loss function -http://leansixsigmadefinition.com/glossary/taguchi-loss-function/
22. Total quality control-http://www.process-improvement-japan.com/total-quality-control.Html
23. Total Quality Management, Bester field, Dale H.,3rd Edition, Pearson Education,
24. Total quality management-https://en.wikipedia.org/wiki/Total_quality_management
25. Walter A Shewhart - https://en.wikipedia.org/wiki/Walter_A._Shewhart
26. Zero defect-https://en.wikipedia.org/wiki/Zero_Defects

ISO 9000
Family and Related Standards

OVERVIEW

ISO 9000 Family of Quality Standards enables an organization to develop a quality management system. The purpose of these standards is to provide management with an understanding of how to utilize all standards appropriately for implementing quality management system in their organization. Prior to introduction of ISO 9000: Quality Managment System—Fundamental and Vocabulary Standard, ISO 8402: Quality Management and Quality Assurance Vocabulary Standard was used. Current edition of standards are: ISO 9000:2015—QMS Fundamentals and Vocabulary, ISO 9001:2015—QMS Requirements and ISO 9004:2018—Quality Management—Quality of an Organization—Guidance to achieve sustained success.

ISO 9000 standard discusses definitions and terminology used to clarify the concepts used by ISO 9001 and ISO 9004 standards. Present ISO 9001 QMS is a business management tool which creates a way of doing business in present day context. The ISO 9001:2015 management system is applicable to each and every aspect of the organization, i.e. from understanding client's requirement to delivery and post-delivery activities. The organization's efficiency, output and quality are aligned and executed optimally by use of ISO 9001 model. It provides a structure through which the management can manage the resources and output of its business in an optimum manner as also monitor quality. The ISO9001 business model outlines the organization's vision, mission, strategies, infrastructure, organizational structure and operational procedures and supports the delivery of product and service through application of effective and continually improving systems, whilst enhancing customer satisfaction.

ISO's Annex SL directives were circulated during 2012. It is a framework which provides guidance that how future ISO Management System Standards (MSS) should be written. The development of Annex SL has been considered as very important event in the history of ISO 9001. The aim of Annex SL is to enhance the consistency and alignment of MSS by providing a unifying and specified high level structure, identical core text and common terms and core definitions. This directive states that all management system standards will use a consistent structure, common text and terminology and this is enacted through high level structure, identical core text, common terms and core definitions.

On account of Annex SL directives, ISO 9001:2015 is consistent in structure with other management system standards (MSS) such as ISO14001:2015—Environment Management System, ISO 45001:2018—Occupational Health and Safety Management System Standard. These standards are covered in this chapter.

3.1 ISO 9000:2015—QUALITY MANAGEMENT SYSTEM—FUNDAMENTALS AND VOCABULARY

3.1.1 ISO 9000:2015 standard provides the foundation for other related quality management standards like ISO 9001:2015 and ISO 9004:2018 and covers following universally applicable aspects related to quality management:

   a. Fundamental concepts of business organizations to meet challenges arising due to present/on-going business environment.

   b. The seven principles which are fundamental beliefs, rules and values accepted as true.

   c. Terms and definitions enabling organizations and interested parties in improving their communication through a clear understanding of the vocabulary. Besides, many new terms included in this standard and the definitions of some old terms have been revised.

The concepts and principles of this standard should be seen in a holistic way so that the organizational QMS model can be developed by using the fundamental concepts and principles of quality management. A formal QMS provides a framework for planning, executing, monitoring and improving the performance of quality management activities. It is important for an organization to regularly evaluate and monitor the implementation of the plan and performance of an organization. The graphical view of development of ISO 9000, ISO 9001 and ISO 9004 standard is shown in Fig. 3.1.

3.1.2 ISO 9000:2015 standard is beneficial to trainers and professionals who give advice on quality management to customers and interested parties seeking confidence in business organization's ability as also, organizations seeking confidence in their supply chain in conformity with the requirements of ISO 9001. This standard proposes QMS framework that integrates established fundamental concepts, principles, processes and resources related to quality. The aim is to increase an organization's awareness of its duties and commitment in fulfilling the needs and expectations of its customers and interested parties.

3.1.3 Quality management fundamental concepts: The fundamental concepts described in ISO 9000:2015 are linked to quality, quality management system, context of organization, and interested parties as also support to QMS like people, competence, awareness and communication. An organization focused on quality promotes a culture that results in the behavior, attitudes, activities and processes that deliver value through fulfilling the needs and expectations of customers and other relevant interested parties. All fundamental concepts have been considered in requirements of ISO 9001:2015 standard and it is noteworthy that fundamental concepts are explicitly covered under various clauses.

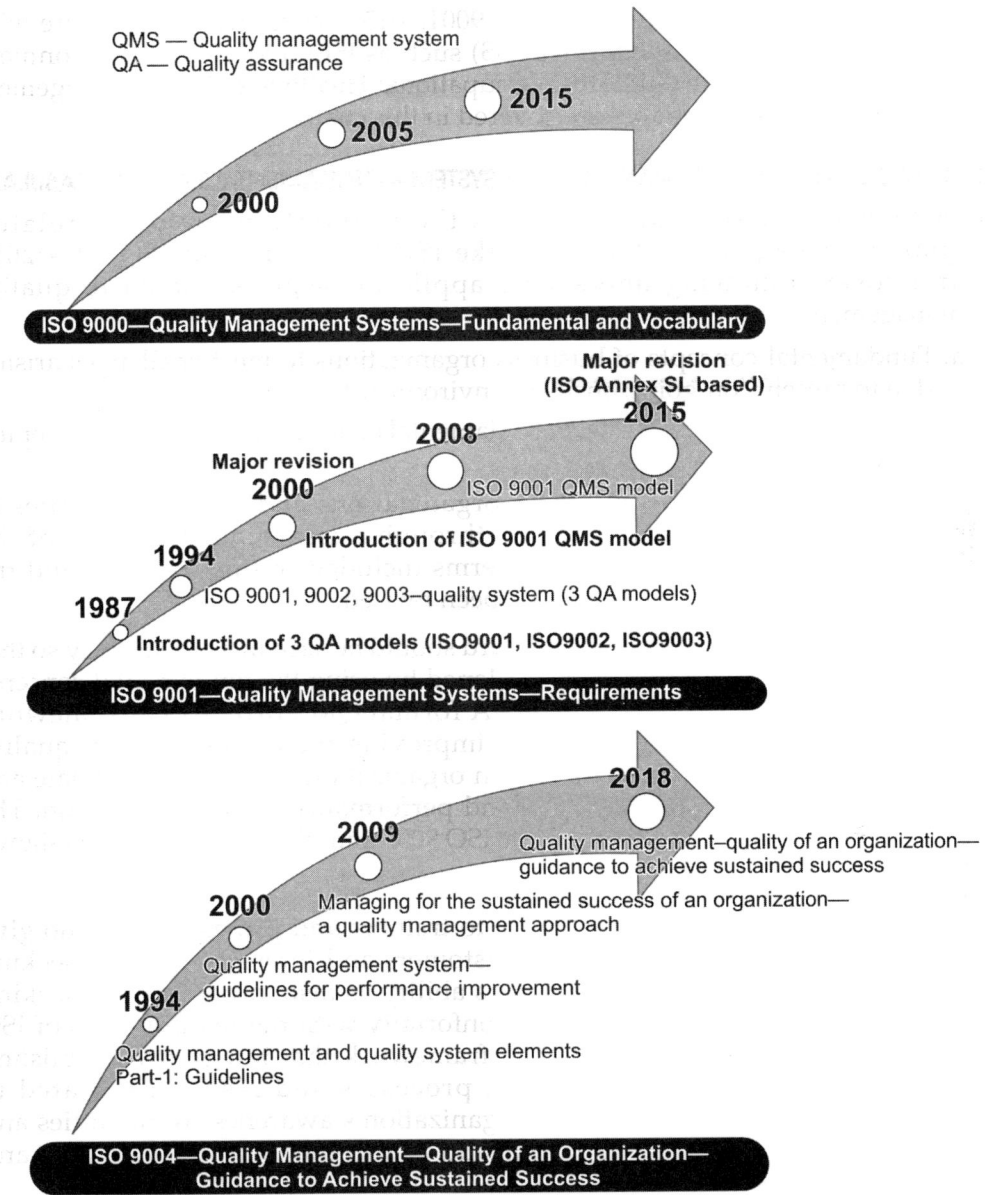

Fig. 3.1: Graphical view of Development of ISO 9000, ISO 9001 and ISO 9004 standards

3.1.4 Quality management principles: ISO 9000:2015 contains seven quality management principles supporting the fundamental concepts and can be used by senior management to promote improvement in the organization. These principles are: (a) Customer focus, (b) leadership, (c) engagement of people, (d) process approach, (e) improvement, (f) evidence-based decision making, (g) relationship management.

 a. Customer focus: "The primary focus of quality management is to meet customer requirements and to strive to exceed customer expectations".

Sustained success is achieved when an organization attracts and retains the confidence of customers and other relevant interested parties. Every aspect of customer interaction provides an opportunity to create more value for the customer. Understanding current and future needs of the customers and other relevant interested parties contributes to sustained success of the organization. It is critical to understand the purpose of quality management and how an organization can attain long-term success by using this principle.

b. *Leadership:* "Leaders at all levels establish unity of purpose and direction and create conditions in which people are engaged in achieving the organization's objectives". Creation of unity of purpose and direction and engagement of people enable an organization to align its strategies, policies, processes and resources to achieve its objectives. It is noteworthy that this principle emphasises on leadership role to be performed by all levels of leaders and make them accountable for engagement of people. The role of leadership is significant because of on-going highly dynamic business environment.

c. *Engagement of people:* "Competent, empowered and engaged people at all levels throughout the organization are essential to enhance the organization's capability to create and deliver value". To manage an organization effectively and efficiently. It is important to involve all people at all levels and to respect them as individuals. Recognition, empowerment and enhancement of competence facilitates the engagement of people in achieving the organization's objectives. The underlying critical aspect which facilitates the creation of value in products and services of an organization is the people of that organization. The success of an organization predominantly depends on its people.

d. *Process approach:* "Consistent and predictable results are achieved more effectively and efficiently when activities are understood and managed as interrelated processes that function as a coherent system". The quality management system consists of interrelated processes. Understanding how results are produced by this system enables an organization to optimize the system and its performance. It is critical to understand the essence of consistent and predictable results of a process and way to optimize the system performance.

e. *Improvement:* "Successful organizations have an on-going focus on improvement". Improvement is essential for an organization to maintain current levels of performance, to react to changes in its internal and external conditions as also to create new opportunities. Improvement and success are closely aligned and create new opportunities to grow.

f. *Evidence-based decision making:* "Decisions based on the analysis and evaluation of data and information are more likely to produce desired results". Decision making can be a complex process, and it always involves some uncertainty. It often involves multiple types and sources of inputs, as well as their interpretation, which can be subjective. It is important to understand the cause and effect relationships and potential unintended consequences. Facts, evidence and data analysis lead to greater objectivity and confidence in decision making. To produce desired results greater objectivity and confidence is required in decision making.

g. *Relationship management:* "For sustained success, organizations manage their relationships with relevant interested parties, such as suppliers". Interested parties influence the performance of an organization. Sustained success is more likely to be achieved when the organization manages relationships with all of its interested parties to optimize their impact on its performance. Relationship management with its supplier and partner network is of particular importance. It is critical to understand the significance of relevant interested parties for sustained success of an organization.

3.1.5 **Concept relationships and their graphical representations:** The terms and definitions belong to different concept classes in ISO 9000:2015 standard. Annex SL core terms and definitions are common to other ISO management system Standards are also included in this standard. There are total thirteen concept classes in the standard including, person or people, organization, activity, process, system, requirement, result, data, information and document, customer, characteristic, determination, action and audit. The graphical representation of each concept class is given in the standard and relationship among related terms is also depicted. Three categories of relationship exist between QMS terms: Generic, partitive and associative relation. Each relation is depicted in a different way: (a) Generic relation by "fan or tree diagram", (b) partitive relation by "rake without arrows line", (c) associative relation by "a line with arrowheads at each end".

In case of generic and partitive relations, the relationship between concepts are based on the hierarchial formation of the characteristic of a species so that most economical description of a concept is formed by naming its species and describing the characteristics that distinguish it from its parents or siblings. Generic relation exists between terms when subordinate concept within the hierarchy inherits all the characteristics of the superordinate concept and contains descriptions of these characteristics which distinguish them from the superordinate (parent) and coordinate (siblings) concepts. Partitive relation exists when subordinate concepts within hierarchy form constituent parts of the superordinate concepts. The associative relations are helpful in identifying the nature of the relationship between one concept and another within a concept class, e.g. cause and effect, activity and location, activity and result, tool and function, material and product. The relation between the terms of the concept class is important to understand in order to use ISO 9001:2015 standard properly. Fig. 3.2 gives graphical representation of concept relation.

a. **Concepts of the class "person or people" and related concepts**
This is a new concept class introduced in ISO 9000:2015 standard. Top management term is one of the core terms and definition of Annex SL. All other terms are new in this concept class. Total 3 generic and 3 associative relations exists between terms of concept class as given in Fig. 3.3.

b. **Concept of the class "organization" and related concepts**
The definition of organization and interested party are as per Annex SL terms and definitions. Some other new terms were introduced like context of

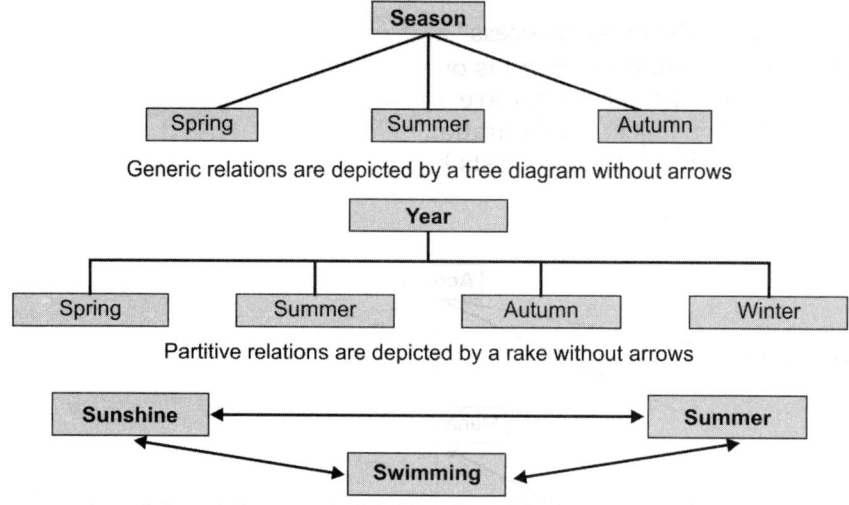

Fig. 3.2: Graphical representation of concept relation

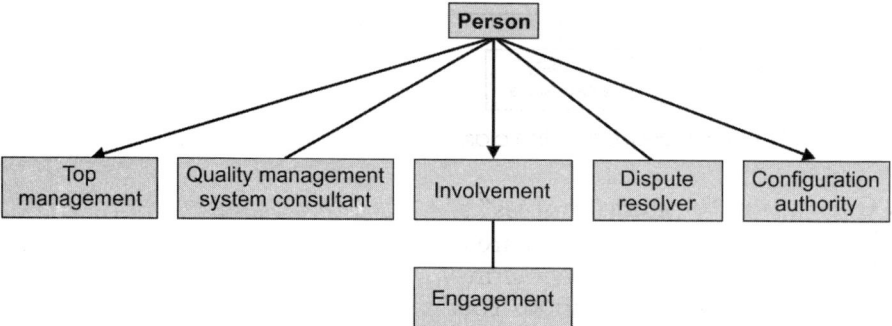

Fig. 3.3: Concepts of the class "person or people" and related concepts

organization, provider, external provider, dispute resolution process (DRP) provider and association. Total 3 generic and 5 associative relations exist between terms of concept class, as given in Fig. 3.4.

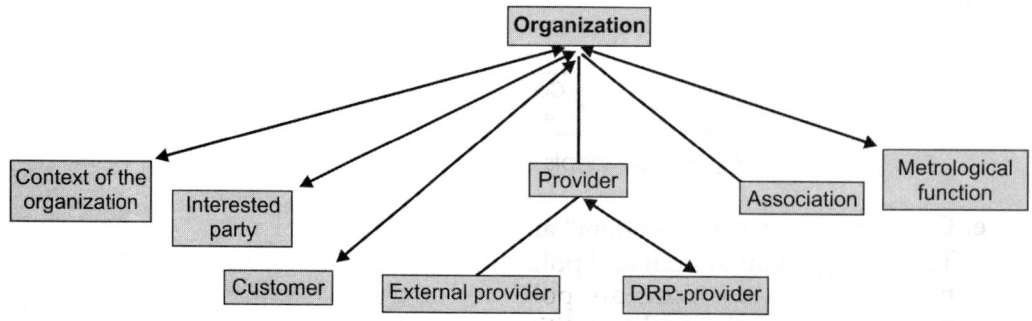

Fig. 3.4: Concepts of the class "organization" and related concepts

c. Concepts of the class "process" and related concepts

The continual improvement is one of the Annex SL term and definition. The newly introduced terms are improvement, activity, change control, configuration object, configuration management. Total 4 generic, 5 associative and 4 portative relations exists between terms of concept class, as given in Fig. 3.5.

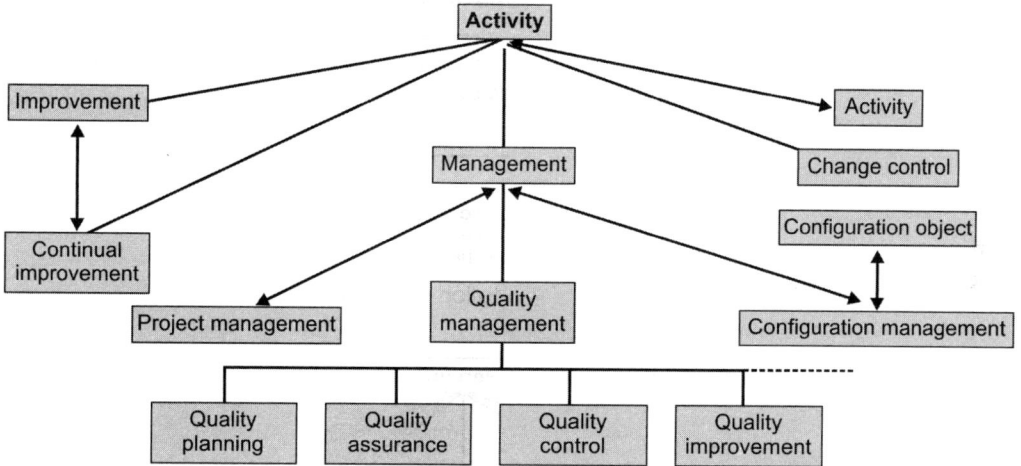

Fig. 3.5: Concepts of the class "activity" and related concepts

d. Concepts of the class "process" and related concepts

The process and outsourced are new as per Annex SL terms and definitions. The newly introduced other terms are quality management system realization and competence acquisition. Total 3 generic and 4 associative relationships exist between terms of concept class, as given in Fig. 3.6.

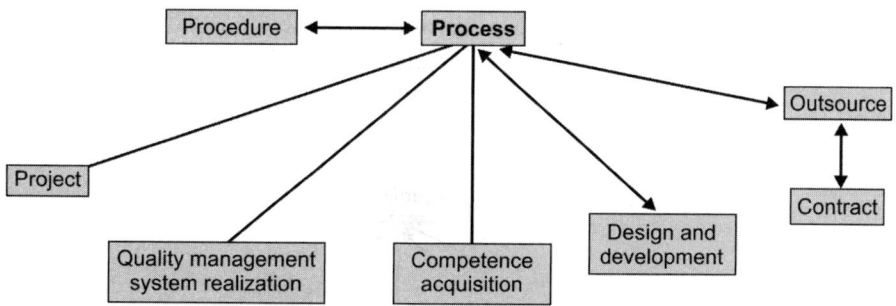

Fig. 3.6: Concepts of the class "process"

e. Concepts of the class "system" and related concepts

The management system and policy are Annex SL terms and definitions. The newly introduced terms are policy vision, mission and strategy. Total 3 generic, 8 associative and 1 partitive relations exists between terms of concept class as given in Fig. 3.7

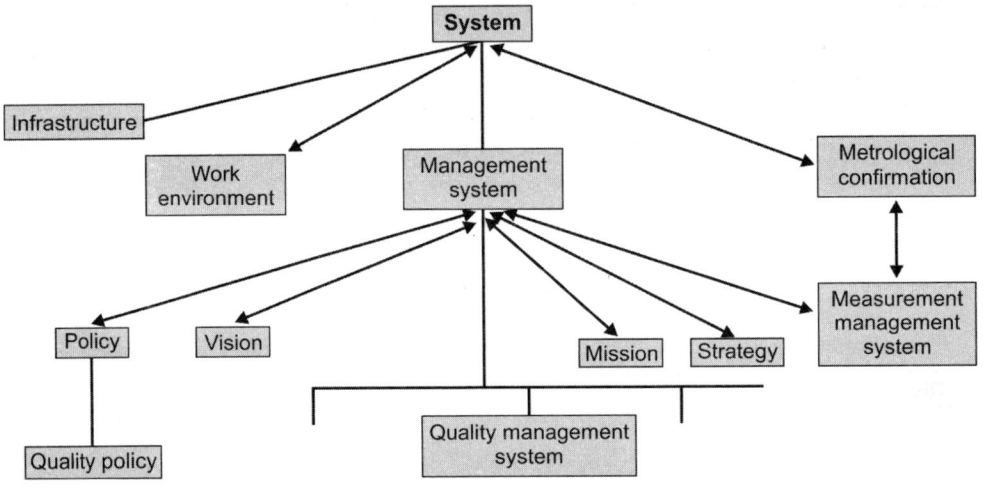

Fig. 3.7: Concepts of the class "system" and related concepts

f. Concepts of the class "requirement" and related concepts

The requirement, conformity and non-conformity are Annex SL terms and definitions. The newly introduced concepts are innovation, quality requirement, statutory requirement, regulatory requirement and product configuration information. Total 6 generic and 10 associative relations exist between terms of concept class, as given in Fig. 3.8.

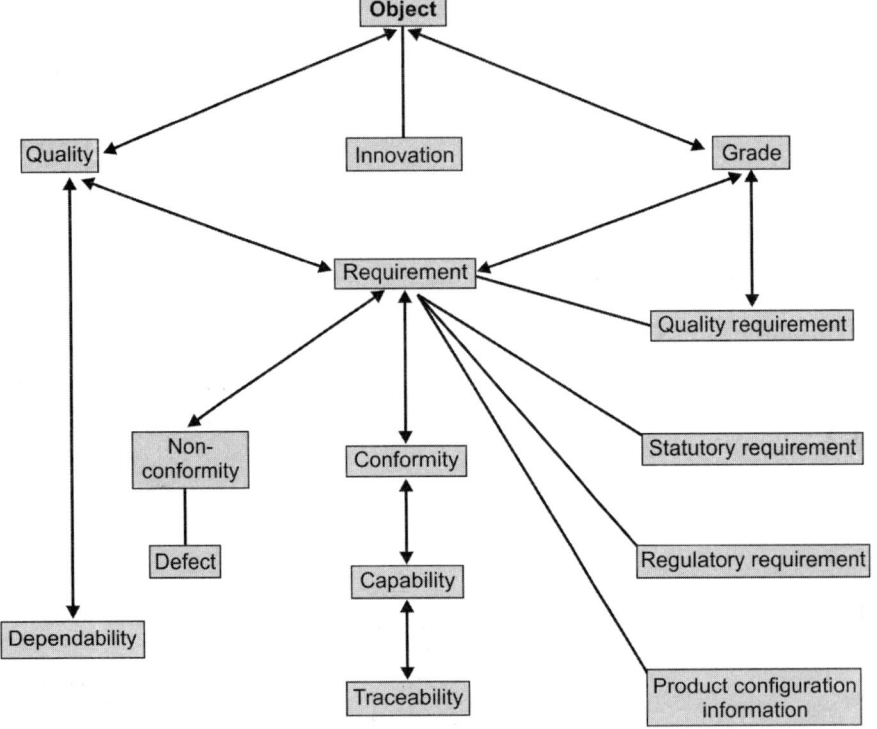

Fig. 3.8: Concepts of the class "requirement" and related concepts

g. **Concepts of the class "result" and related concepts**

The risk, performance, objective and effectiveness are Annex SL terms and definitions. The newly introduced relations are risk, performance, output, service, objective, success and sustained success. Total 7 generic and 4 associative relations exist between terms of concept class, as given in Fig. 3.9.

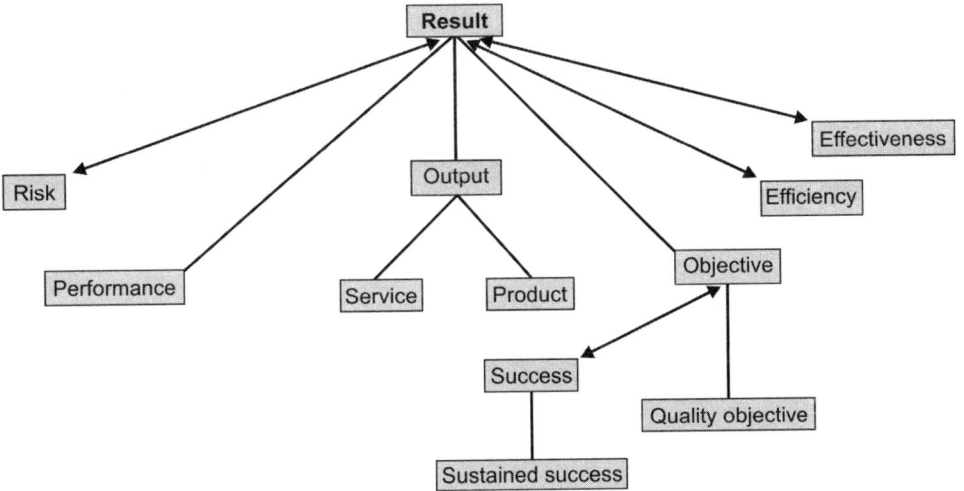

Fig. 3.9: Concepts of the class "result" and related concepts

h. **Concepts of the class "data, information and document", and related concepts**

The documented information is one of the Annex SL term and definition. The newly introduced relations are documented information, information system, project management plan, configuration status accounting and specific plan. Total 8 generic and 6 associative relation exists between terms of concept class, as given in Fig. 3.10.

i. **Concepts of the class "customer" and related concepts**

This is a new concept class having 5 new terms and definitions (feedback, complaint, dispute, customer service and customer satisfaction code of conduct). Total 9 associative relation exists between term of concept class, as given in Fig. 3.11.

j. **Concepts of the class "characteristic" and related concepts**

The competence is one of the Annex SL terms and definition. The newly introduced concepts are human factor, configuration and configuration baseline. Total 3 generic and 3 associative relations exist between terms of concept class, as given in Fig. 3.12.

k. **Concepts of the class "determination" and related concepts**

The monitoring and measurement are Annex SL terms and definitions. The newly introduced concepts are monitoring, measurement and performance evaluation. Total 3 generic and 5 associative relations exists between terms of concept class, as given in Fig. 3.13.

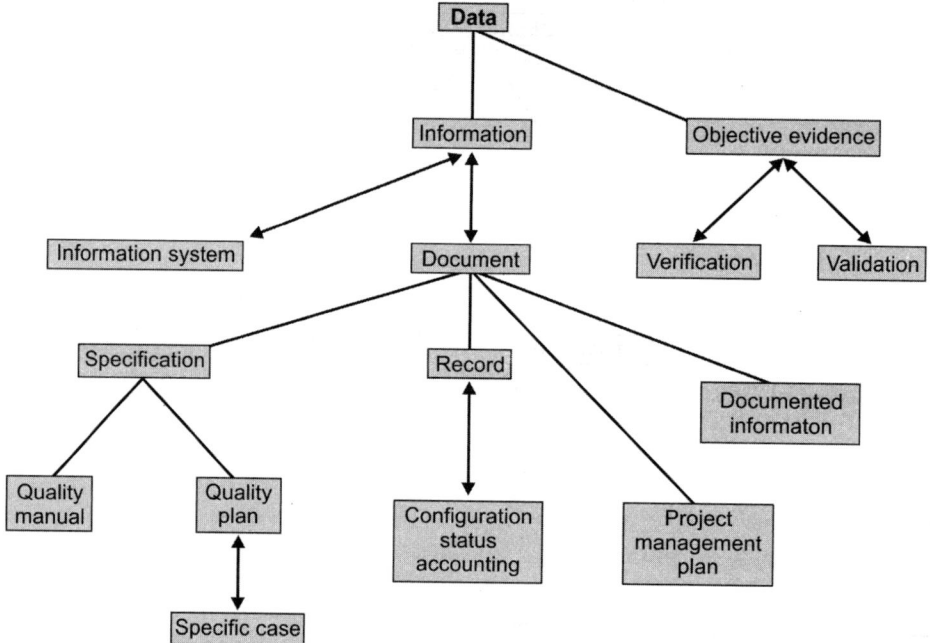

Fig. 3.10: Concepts of the class "data, information and document" and related concepts

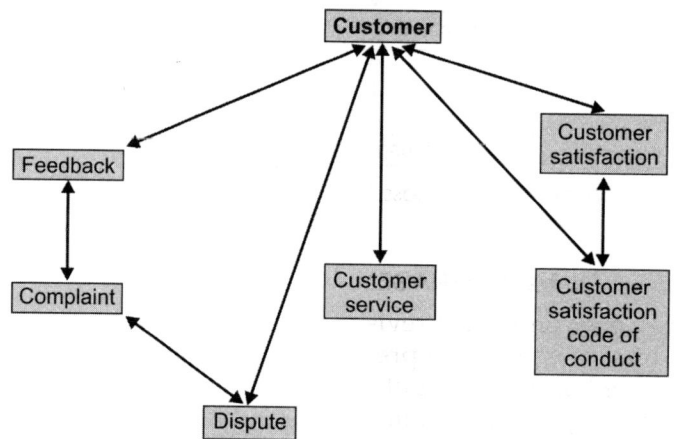

Fig. 3.11: Concepts of the class "customer" and related concepts

l. **Concepts of the class "action" and related concepts**

The corrective action is one of the Annex SL term and definitions. Total 9 generic and 2 associative relations exist between terms of concept class, as given in Fig. 3.14.

m. **Concepts of the class "audit" and related concepts**

The audit is one of the Annex SL terms and definition. The newly introduced relations are joint audit, combined audit, guide and observer. Total 2 generic and 16 associative relations exist between terms of concept class, as given in Fig. 3.15.

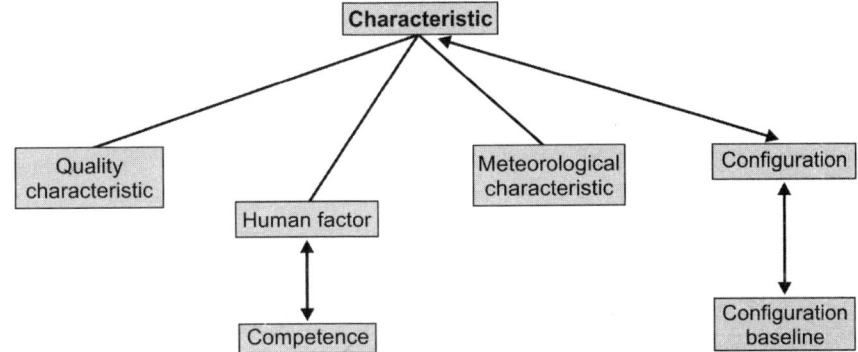

Fig. 3.12: Concepts of the class "characteristic" and related concepts

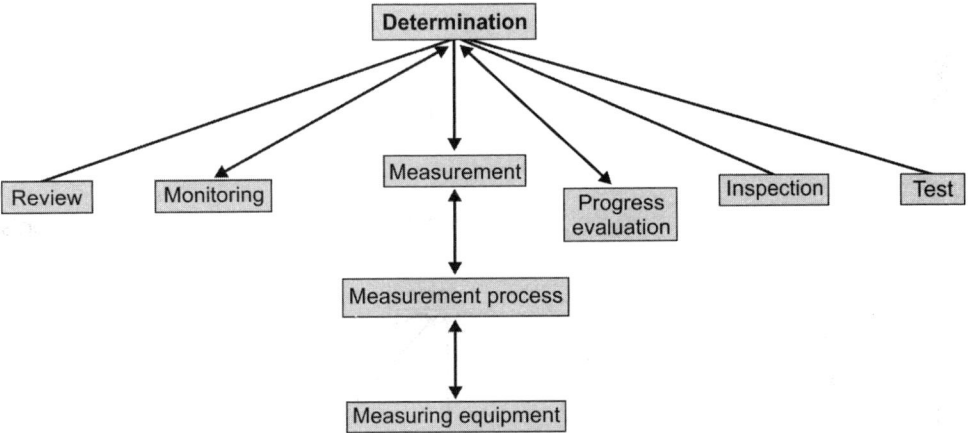

Fig. 3.13: Concepts of the class "determination" and related concepts

3.2 ISO 9001:2015—QUALITY MANAGEMENT—REQUIREMENTS

3.2.1 ISO 9001:2015 is second major revision of ISO 9001 standard. The 2015 version is also less prescriptive than its predecessors and focuses on performance. The standard is based on 10 clauses of high-level structure and can be better aligned with other management system standards of disciplines like, environment, health and safety, etc. These ten clauses are: Scope, normative reference, terms and definitions, context of organization, leadership, planning, support, operation, performance evaluation and improvement.

3.2.2 According to the standard top management of an organization is required to be more involved and accountable for quality which is aligned with wider business strategy. Risk-based thinking throughout the standard makes the whole management system a preventive tool and encourages continuous improvement.

3.2.3 To achieve intended results consistently along with effectiveness and efficiency, the process approach is of significant importance. The standard promotes adoption of process approach. This is achieved by combining the process

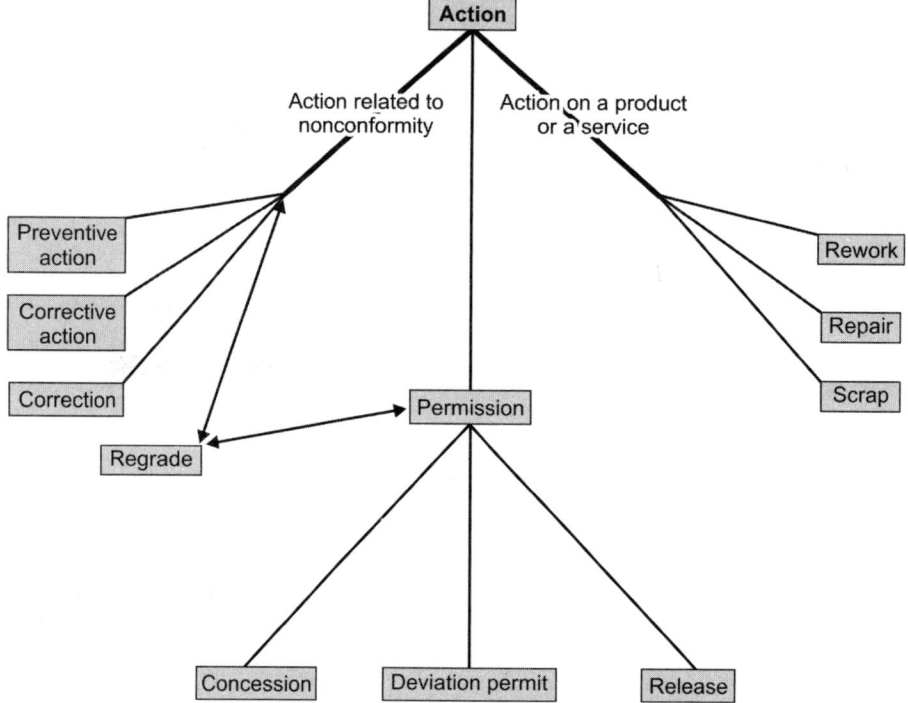

Fig. 3.14: Concepts of the class "action" and related concepts

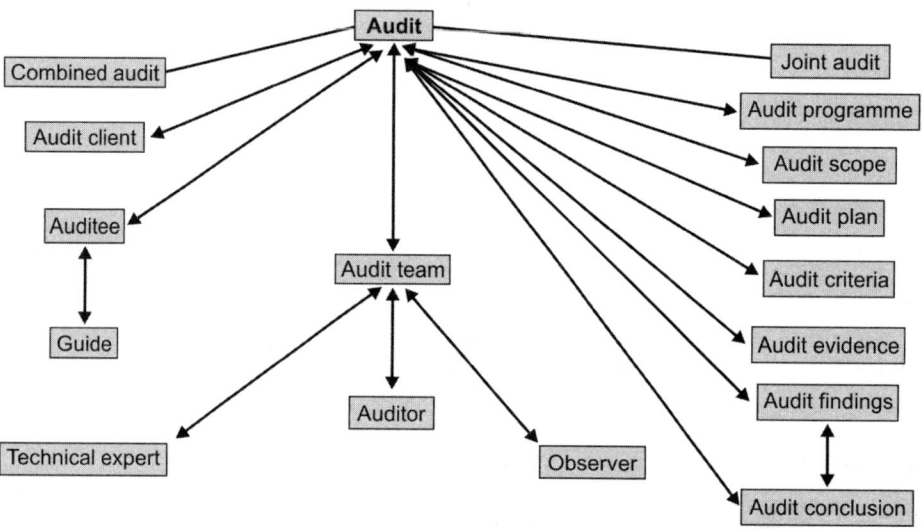

Fig. 3.15: Concepts of the class "audit" and related concepts

approach with risk-based thinking, and employing the Plan-Do-Check-Act cycle at all levels in the organization.

3.2.4 It is important to understand that merely customer requirement is not the basis of ISO 9001:2015 process-based QMS model. Great emphasis has been given on understanding context of the organization. This is achieved by monitoring and reviewing the impact of organizational internal issues and external issues on QMS. Additionally, identifying needs and expectations of relevant interested parties have also been given more impetus.

3.2.5 This process model also gives higher significance on achievement of intended results, which in turn results in intended product and services along with higher customer satisfaction. Leadership is now explicit requirement for improvement cycle of ISO 9001:2015 model and is essential for the effectiveness of clauses pertaining to planning, support, operation, performance evaluation and improvement. The "Plan-Do-Check-Act" (PDCA) cycle as given in Fig. 3.16 can be briefly described as follows:

a. **Plan:** Establish the objectives of the system, its processes and the resources needed to deliver results in accordance with customers' requirements and the organization's policies. Identify and address risks and opportunities.

b. **Do:** Implement what was planned

c. **Check:** Monitor and (where applicable) measure, processes, the resulting products and services against policies, objectives, requirements and planned activities. Report the results;

d. **Act:** Take actions to improve performance, as necessary.

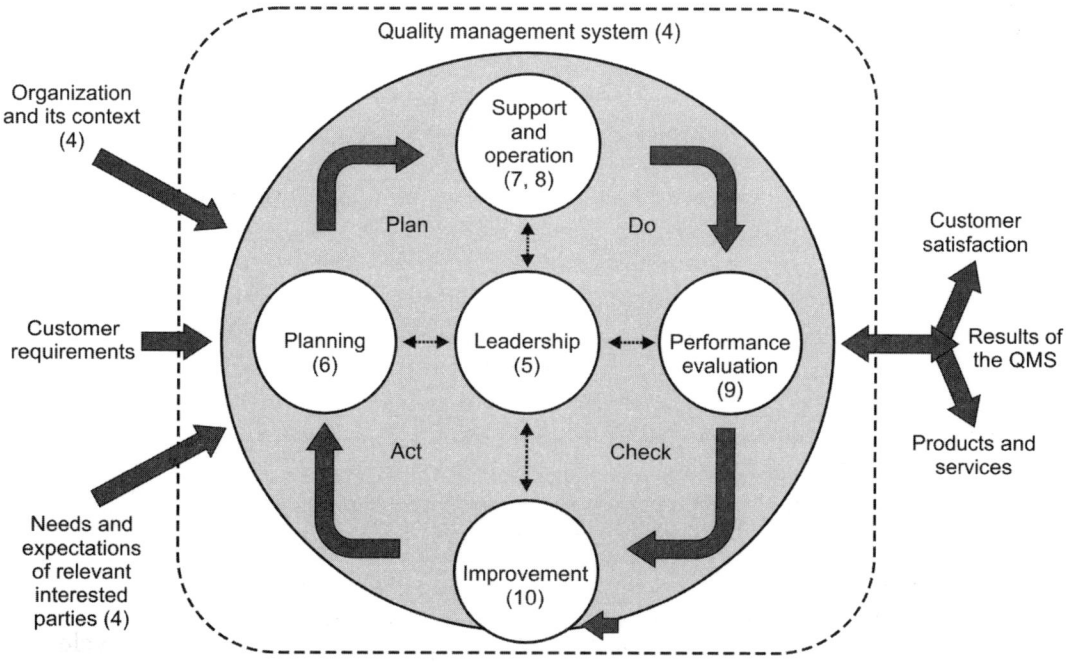

Fig. 3.16: Plan-Do-Check-Act cycle

ISO 9001:2015 Clause 6—planning is related to "Plan", Clause 7—support, and Clause 8—operation are related to "Do", Clause 9—performance evaluation is related to "Check" and Clause 10—improvement is related to "Act" of PDCA cycle. Clause 4—organization and its context, needs and expectation of relevant interested parties is significant in development and review of QMS and Clause 5—leadership is important for each stage of Plan-Do-Check-Act cycle.

3.3 ISO 9004:2018—QUALITY MANAGEMENT—QUALITY OF AN ORGANIZATION: GUIDANCE TO ACHIEVE SUSTAINED SUCCESS

3.3.1 ISO 9004: 2018 standard gives guidance to achieve sustained success. It acts as a guide to organizations to extend the benefits of ISO 9001 and develop their performance through continual improvement. The standard offers guidance as to how organizations can enhance their overall quality by improving their maturity level, and provides a framework for strategy, leadership, resources and processes. The standard is aligned with the concepts and terminology of ISO 9000:2015 and ISO 9001:2015. This standard is not certifiable standard like ISO 9001. The standard assists organizations to identify and balance the needs and expectations of their customers with those of other interested parties. Improvement, learning and innovation are key elements which support sustained success. The standard promotes the systematic improvement of the organization's overall performance in a complex, demanding, ever-changing and evolving business environment.

3.3.2 ISO 9004:2018 standard focuses on the concepts related to "quality of an organization", and "identity of an organization". (a) Quality of the organization—in order to achieve sustained success, the degree to which the inherent characteristics of the organization fulfill the needs and expectations of its customers and other interested parties. (b) The identity of the organization is determined by its characteristics, based on its mission, vision, values and culture. ISO 9004:2018 helps as self-assessment tool to understand the level of maturity of the organization's management system. The results of self-assessment can be a valuable input into the management review process. The structure of ISO 9004:2018 standard is represented in Fig. 3.17. It also explains that the standard is helpful to an organization to enhance confidence in the organization's ability to meet customer focus and needs and expectations of other relevant interested parties.

3.3.3 ISO 9004 encourages organisations to go beyond the quality of their products and services and the needs and expectations of their customers. To achieve sustained success, organisations should focus on anticipating and exceeding the needs and expectations of their interested parties, with the purpose of enhancing their overall performance.

Factors affecting an organization's success continually evolve over the years and adapting to these changes is important for sustained success.

Clause 4 of ISO 9004:2018 states that the quality of the organization is the degree to which the inherent characteristics of the organization fulfill the needs and expectations of its customers and other interested parties, in order to achieve sustained success.

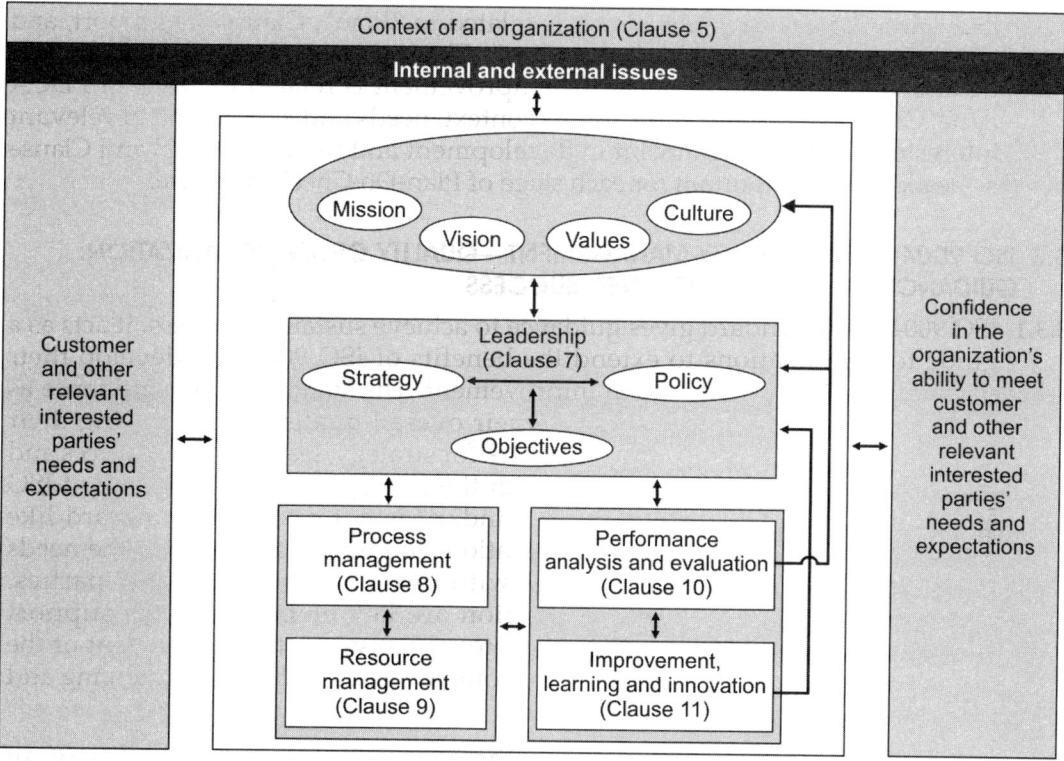

Fig. 3.17: Representation of the structure of ISO 9004:2018 standard

Clause 5 identifies three key factors to be considered when determining the context of the organisation as follows:

• Interested parties;

• External issues such as statutory and regulatory requirements, competition, globalisation, social, economic, political and social factors, innovation and advances in technology;

• Internal issues such as size and complexity of the organisation, resources available or lacking, levels of competence and organisational knowledge, maturity and innovation.

The ability to achieve sustained success is enhanced by organisational personnel at all levels comprehending the organization's evolving context. Improvement and innovation also support sustained success.

Clause 6 states that the identity of the organization is determined by its characteristics, based on its mission, vision, values and culture.

Clause 7 addresses leadership issue.

Clause 8 on process management reinforces the thesis that consistent and predictable results are achieved more effectively and efficiently when the network of processes function as a coherent system.

Clause 9 requires the organization to determine and manage the resources needed to achieve its objectives.

Clause 10 requires the organization to establish a systematic approach to the collection, analysis and review of available information. Based on the results, the organization should use the information to update its understanding of its context, policies, strategy and objectives, while also promoting improvement, learning and innovation activities.

Clause 11 states that improvement, learning and innovation are interdependent and it is the key aspects that contribute to the sustained success of an organization.

3.4 ISO 14001:2015—ENVIRONMENTAL MANAGEMENT SYSTEM

3.4.1 ISO 14001:2015 specifies the requirements for an environmental management system that an organization can use to enhance its environmental performance through more efficient use of resources and reduction of waste, gaining a competitive advantage and the trust of stakeholders. ISO 14001:2015 is intended for use by an organization seeking to manage its environmental responsibilities in a systematic manner that contributes to the environmental pillar of sustainability. An environmental management system (EMS) contributes to reducing an organization's impact on the environment while improving operating efficiency. EMS aids an organization to comply with environmental regulations, and improve health and safety for both employees and the community.

3.4.2 ISO 14001:2015 standard helps organizations to monitor and control their environmental issues in a "holistic" manner, facilitates improvement in environmental performance. The ISO 14001:2015 places higher importance of environmental management within the organization's strategic planning and requires suitable dealing with the issues of leadership, commitment and stakeholder-focused communication strategy. This standard can be easily integrated with ISO 9001:2015—Quality Management System Standard and ISO 45001-Occupational Health and Safety Standard.

3.4.3 ISO 14001:2015 is applicable to any organization, regardless of types and sizes, be they private, not-for-profit or governmental. It requires that an organization considers all environmental issues relevant to its operations, such as air pollution, water and sewage issues, waste management, soil contamination, climate change mitigation and adaptation, as also resource use and efficiency. Like all ISO management system standards, ISO 14001 includes the need for continual improvement of an organization's systems and approach to environmental concerns. In revised version there are key improvements such as the increased prominence of environmental management within the organization's strategic planning processes, greater input from leadership and a stronger commitment to proactive initiatives that boost environmental performance.

3.4.4 The organization shall plan to take actions to address its environmental concerns, risks and opportunities, and compliance obligations. ISO 14001:2015 emphasizes that an organization has to consider a life cycle perspective-which includes carriage, disposal, and recycling as well as production—all environmental aspects of its products, services, and activities that are deemed to be within the organization's control. Changes or planned future changes to services also have to be taken into account.

There are many reasons why an organization should take a strategic approach to improving its environmental performance. ISO 14001:2015 helps: (a) Demonstrate compliance with current and future statutory and regulatory requirements; (b) Increase leadership involvement and engagement of employees, and (c) Improve company reputation and the confidence of stakeholders through strategic communication

3.4.5 ISO 14001:2015 respond to the latest trends, including the increasing recognition by companies of the need to factor in both external and internal elements that influence their environmental impact, such as climate volatility and the competitive context in which they work. The changes also ensure that the standard is compatible with other management system standards.

3.4.6 ISO 14001:2015 now requires: (a) Environmental management to be more prominent within the organization's strategic direction, (b) a greater commitment from leadership, (c) the implementation of proactive initiatives to protect the environment from harm and degradation, such as sustainable resource use and climate change mitigation, and (d) a focus on life-cycle thinking to ensure consideration of environmental aspects from development to end-of-life.

3.4.7 ISO 14001:2015 successful implementations demonstrate assurance to interested parties that an appropriate environmental management system is in place. EMS generally varies from organization to organization based on factors such as size, compliance obligations, sector worked in, past history and performance.

3.5 ISO 45001:2018—OCCUPATIONAL HEALTH AND SAFETY MANAGEMENT

3.5.1 ISO 45001:2018 is a new International Standard on occupational health and safety (OH&S), providing a framework for managing the prevention of injury, work-related injury and ill health, with the intended outcome of improving and providing a safe and healthy workplace for workers. This standard enables organization to proactively improve its OH&S performance in preventing injury and ill-health and facilitates in addressing other related aspects such as worker wellness/wellbeing. Simultaneously reduction in incidents of disruption to operations, absenteeism, employee turnover, etc. can also be achieved. It does not address issues such as product safety, property damage or environmental impact, beyond the risk to workers and other relevant interested parties. It does not address issues such as product safety, property damage or environmental impacts, beyond the risks to workers and other relevant interested parties.

3.5.2 ISO 45001:2018 has been developed by considering relevant guidelines of International Labour Organization, National standards as well as BS OHSAS 18001 standard. Similar to ISO 9001:2015 and ISO 14001:2015, the implementation of an OH&S management system will be a strategic decision for an organization to support its sustainability initiatives and ensuring that people are safer and healthier. This standard also follows the ten clauses high level structure approach that is being applied to other ISO management system standards, such as ISO 9001:2015 and ISO 14001:2015. This standard emphasizes the need for worker participation in the functioning of an occupational health and safety (OH&S) management system, as well as requiring that an organization ensures that its workers are competent to do their assigned tasks safely. It is intended to be applicable to any organization regardless of its size, type and nature. All of its requirements are intended to be integrated into an organization's own management processes.

3.5.3 ISO 45001:2018 can be used in whole or in part to systematically improve occupational health and safety management. However, claims of conformity to this document are not acceptable unless all its requirements are incorporated into an organization's OH&S management system and fulfilled without exclusion. It uses simple Plan-Do-Check-Act (PDCA) model, which provides a framework for organizations to plan what they need to put in place, in order to minimize the shortcomings. The measures should address concerns that can deal with long-term health issues and absence from work, as well as those that give rise to accidents. The standard calls for the organization's management and leadership to integrate responsibility for health and safety issues as part of organization's overall plan.

3.5.4 According to ISO 45001:2018 the result of context review is to be used for: (a) Determining scope and issues (positive and negative) that can affect how organization manages OH&S management system, (b) determining risk and opportunities, (c) developing or enhancing OH&S policy and set objectives, (d) high level understanding needs and expectations of workers (i.e. both managerial and non-managerial workers) and other interested parties. Understanding needs and expectations of interested parties implies needs and expectations of both managerial and non-managerial workers and worker's representatives, occupational health and safety professionals (e.g. doctor, nurses, etc.). Leadership clause has enhanced commitment and active support from top management. The leader has to take overall responsibility and accountability for protection of worker's work-related health and safety. More emphasis is required on non-managerial worker participation, identification for needs for competence, training, etc.

3.5.5 According to ISO 45001:2018, hazard identification should proactively identify any source or situation arising from organizations activities, with potential for work-related injury and ill-health. When planning for OH&S management system, organization shall refer to "organizational context" requirement referred to in "Interested parties" as also scope of its OH&S management system and determine risks and opportunities that need to be addressed. Organizations should maintain and retain documented information of OH&S objectives and plan to achieve them, keeping complexity to a minimum. Organizations should ensure that outsourced processes affecting OH&S management system are controlled. It is necessary to establish the controls to ensure that the procurement of goods and services conforms to its OH&S management system requirements.

Organization shall also establish, implement and maintain a process for monitoring, measurement and evaluation in conformity with the standard.

3.6 ISO AND CONSTRUCTION

ISO has published a brochure on "ISO and Construction". This brochure gives a concise overview of ISO's substantial portfolio of International standards for the construction sector. It underlines how ISO standards tackle the challenges of sustainable development at the same time as providing requirements for technical and functional performance. ISO standards improve safety, sustainability and durability in construction.

Why do we Need ISO Standards for Construction?

The world's rapid population growth and rampant urbanization have brought an increasing need for a high-quality, safe and sustainable built environment. In the world of building and construction, ISO standards help codify international best practice and technical requirements to ensure buildings and other structures (known as civil engineering works) are safe and fit for purpose.

Updated on a regular basis to account for climate, demographic and social changes, ISO's standards for construction are developed with input from all stakeholders involved, including architects, designers, engineers, contractors, owners, product manufacturers, regulators, policy makers and consumers.

What Standards does ISO have for Construction?

Important ISO standards related to buildings and construction cover the important elements like structures; masonry; building materials and products; information management in construction; energy performance and sustainability; heating, cooling and lighting; fire safety and firefighting; lifts and escalators; concrete and cement; design life, durability and service life planning, timber.

Structures

Ensuring all the components of structures are strong enough to withstand appropriate loads and everything fits together as it should be the objective of a number of ISO standards for construction. By establishing defined specifications and test methods, they help ensure structures are designed and built to agreed levels of quality.

ISO/TC 98, bases for design of structures, lays down the basic requirements for the design of structures. With standards focusing especially on terminology and symbols, loads and forces, it ensures constructions are built to last and can withstand outside forces such as extreme weather events and natural disasters.

ISO/TC 167, steel and aluminium structures, develops standards that specify requirements for the structural use of steel and aluminium alloys in the design, fabrication and erection of buildings and civil engineering works. Its scope of work includes materials, structural components and connections.

ISO/TC 165, timber structures, deals with the strength and load requirements of structural timber, while geotechnical analysis (interactions between soil and structure) is the focus of ISO/TC 182, geotechnics.

Building Materials and Products

Being able to count on reliable, quality materials is essential for the construction of safe and robust buildings. ISO has more than 100 standards related to the raw materials used in construction, such as concrete, cement, timber and glass. These include standards on terminology, testing procedures and the assessment of safety levels. We also have over 500 standards on building products, such as doors and windows, wood-based panels, floor coverings, ceramic tiles and plastic pipes and fittings. These not only determine the correct dimensions and specifications to ensure products are manufactured to agreed quality levels, but also define test methods for assessing product safety and resistance to things like crushing or chemicals, so that they do not fail or deteriorate prematurely.

Energy Performance and Sustainability

From insulation to energy-using products, improving the energy performance of buildings can make a significant contribution to climate-related targets. As a result, building regulations increasingly require energy-efficient designs and measures are put in place to help improve overall performance.

ISO/TC 163, thermal performance and energy use in the built environment, has more than 130 standards providing guidelines and methods for the calculation of energy consumption in buildings, covering areas such as heating, lighting, ventilation and so forth. ISO's energy standards portfolio includes the recently published series

ISO 52000, energy performance of buildings—overarching EPB assessment, which defines methods to help architects, engineers and regulators assess the overall energy performance of new and existing buildings in a holistic way.

ISO/TC 205, building environment design, has a range of standards defining methods and processes for the design of new buildings and retrofit of existing buildings, to create acceptable indoor environments and practicable energy conservation and efficiency. In addition, we produce standards for measuring the carbon emissions of buildings and others structures, including:

ISO 21930, sustainability in buildings and civil engineering works—core rules for environmental product declarations of construction products and services, which establishes good practices for making environmental claims and communications in the construction sector.

Fire Safety and Fire Fighting

Fires cause destruction and devastation, costing the lives and livelihoods of people. With the increased density of housing, protecting against fires and detecting fire risks have never been more important.

ISO/TC 21, equipment for fire protection and fire fighting, develops standards covering fire protection and fire-fighting apparatus and equipment, including fire extinguishers and fire and smoke detectors.

ISO/TC 92, fire safety, develops standards for assessing fire risks to life and property and mitigating such risks by determining the behaviour of construction materials and building structures.

ISO 7240, fire detection and alarm systems, defines the specifications of fire detection and alarm system equipment used in and around buildings—including their testing and performance—in order to ensure they function effectively.

Information Management in Construction

Since most construction works are project based, having documentation that is clearly understood by all stakeholders is essential to ensure each project is realized in a cost effective manner. Building information models (BIM) are shared digital representations of the physical and functional characteristics of any built object (including buildings, bridges and roads) and form a reliable basis for decision making. They also help protect against the loss of valuable information between stages and processes.

ISO/TC 59/SC 13, organization of information about construction works, develops standards that define the common terms of reference and terminology used in BIMs, as well as requirements for the digital exchange of documentation and data.

Examples include:

ISO 16757-1, data structures for electronic product catalogues for building services.

Part 1: Concepts, architecture and model

ISO/TS 12911, framework for building information modelling (BIM) guidance.

Lifts and Escalators

Rising urbanization and denser populations mean buildings across the world are getting taller. Efficient lifts and escalators are thus essential to cope with the increased loads and access needs and must be operable in times of disaster, such as fire, to evacuate high-rise structures.

ISO/TC 178, Lifts, escalators and moving walks, has over 50 standards, either published or in development, for all kinds of lifts. These cover requirements for everything from planning and installation to energy performance and safety.

One prominent example is:

ISO/TS 18870, lifts (elevators)—requirements for lifts used to assist in building evacuation.

Design Life, Durability and Service Life Planning

ISO/TC 59/SC 14, design life, develops standards that offer a methodology and guidance on how to plan the service life of buildings, including predicting costs and the frequency of maintenance and repairs over their life cycle. The ISO 15686 series on service life planning deals with a wide-range of subjects in this area, such as performance audits and reviews, lifecycle assessment and maintenance and life-cycle costing.

An example is:

ISO 15686-5, buildings and constructed assets—service life planning.

Part 5: Life-cycle costing, which helps track the cost performance over an asset's lifespan.

3.7 OTHER RELEVANT ISO STANDARDS

Several other ISO standards have also been published some of which can be of relevance for quality management system.

A list of these standards is given below:

1. ISO 10001: Quality Management—Customer Satisfaction—Guidelines for Codes of Conduct for Organization
2. ISO 10002: Quality Management—Customer Satisfaction—Guidelines for Complaints Handling in Organizations
3. ISO 10003: Quality Management—Customer Satisfaction—Guidelines for Dispute Resolution External to Organizations
4. ISO 10004: Quality Management—Customer Satisfaction—Guidelines for Monitoring and Measuring
5. ISO 10005: Quality Management—Guidelines for Quality Plans
6. ISO 10006: Quality Management—Guidelines for Quality Management in Projects
7. ISO 10007: Quality Management—Guidelines for Configuration Management
8. ISO 10008: Quality Management—Customer Satisfaction—Guidelines for Business-to-consumer Electronic Commerce Transactions
9. ISO 10012: Measurement Management Systems—Requirements for Measurement Processes and Measuring Equipment
10. ISO 10014: Quality Management—Guidelines for Realizing Financial and Economic Benefits
11. ISO 10015: Quality Management—Guidelines for Training
12. ISO 10018: Quality Management—Guidelines on People Involvement and Competence
13. ISO 10019: Guidelines for the Selection of Quality Management System Consultants and use of Their Services
14. ISO 14040: Environmental Management—Life Cycle Assessment—Principles and Framework
15. ISO 14044: Environmental Management—Life Cycle Assessment—Requirements and Guidelines
16. ISO 19600: Compliance Management Systems—Guidelines
17. ISO 22000: Food Safety Management Systems—Requirement for any Organization in the Food Chain
18. ISO 22316: Security and Resilience—Organizational Resilience—Principles and Attributes
19. ISO 26000: Guidance on Social Responsibility
20. ISO 31000: Risk Management—Guidelines
21. ISO 37001: Anti-bribery Management Systems—Requirements with Guidance for use
22. ISO 39001: Road Traffic Safety (RTS) Management Systems—Requirements with Guidance for use
23. ISO 50001: Energy Management Systems—Requirements with Guidance for use
24. ISO 27000: Information Technology—Security Techniques—Information Security Management Systems—Overview and Vocabulary
25. ISO 27001: Information Technology—Security Techniques—Information Security Management Systems—Requirements

26. ISO 27002: Information Technology—Security Techniques—Code of Practice for Information Security Controls

27. ISO Handbook. ISO 9001:2015 for Small Enterprises—What to Do? 2016. Available at: https://www.iso.org/publication/PUB100406.html

28. ISO. Guidance on the concept and use of the process approach for management systems. ISO/TC 176/SC 2/N 544R3, 2008. Available at: https://www.iso.org/files/live/sites/isoorg/files/archive/pdf/en/04_concept_and_use_of_the_process_approach_for_management_systems.pdf

29. ISO information and guidance on ISO 9001 and ISO 9004. Available at: https://committee.iso.org/tc176sc2

30. ISO 9001: Auditing Practices Group. Various papers. Available at http://committee.iso.org/sites/tc176sc2/home/page/iso-9001-auditing-practices-grou.html

3.8 QMS TERMS AND DEFINITIONS

1. **Audit:** Systematic, independent and documented process for obtaining objective evidence and evaluating it objectively to determine the extent to which the audit criteria are fulfilled.

2. **Competence:** Ability to apply knowledge and skills to achieve intended results.

3. **Complaint:** Expression of dissatisfaction made to an organization, related to its product or service, or the complaints-handling process itself, where a response or resolution is explicitly expected.

4. **Context of organization:** Combination of internal and external issues that can have an effect or an organization's approach to developing and achieving its objectives.

5. **Continual improvement:** Recurring activity to enhance performance.

6. **Corrective action:** Action to eliminate the cause of a nonconformity and to prevent recurrence.

7. **Customer:** Person or organization that could or does receive a product or service that is intended for or required by this person or organization.

8. **Customer satisfaction:** Customer's perception of the degree to which the customer's expectations have been fulfilled.

9. **Defect:** Nonconformity related to an intended or specified use.

10. **Design and development:** Set of processes that transform requirements for an object into more detailed requirements for that object.

11. **Documented information:** Information required to be controlled and maintained by an organization and the medium on which it is contained.

12. **Effectiveness:** Extent to which planned activities are realized and planned results are achieved.

13. **Engagement:** Involvement in, and contribution to, activities to achieve shared objectives.

14. **External provider:** Provider that is not part of the organization.

15. **Feedback:** Opinions, comments and expressions of interest in a product, a service or a complaints—handling process.

16. **Innovation:** New or changed object realizing or redistributing value.

17. **Interested party:** Person or organization that can affect, be affected by, or perceive itself to be affected by a decision or activity.

18. **Management:** Coordinated activities to direct and control an organization.

19. **Management system:** Set of interrelated or interacting elements of an organization to establish policies and objectives, and processes to achieve those objectives.

20. **Measurement:** Process to determine a value

21. **Mission:** Organization's purpose for existing as expressed by top management.

22. **Monitoring:** Determining the status of a system, a process, a product, a service or an activity.

23. **Non-conformity:** Non-fulfillment of a requirement.

24. **Objective:** Result to be achieved.

25. **Organization:** Person or group of people that has its own functions with responsibilities, authorities and relationships to achieve its objectives.

26. **Output:** Result of a process.

27. **Performance:** Measurable result

28. **Provider:** Organization that provides a product or a service.

29. **Policy:** Intentions and direction of an organization as formally expressed by its top management.

30. **Process:** Set of interrelated or interacting activities that use inputs to deliver an intended result.

31. **Product:** Output of an organization that can be produced without any transaction taking place between the organization and the customer.

32. **Project:** Unique process, consisting of a set of coordinated and controlled activities with start and finish dates, undertaken to achieve an objective conforming to specific requirements, including the constraints of time, cost and resources.

33. **QMS consultant:** Person who assists the organization on quality management system realization, giving advice or information.

34. **Quality:** Degree to which a set of inherent characteristics of an object fulfills requirements.

35. **Quality management system:** Part of a management system with regard to quality.

36. **Quality planning:** Part of a management focused on setting quality objectives and specifying necessary operational processes, and related resources to achieve the quality objectives.

37. **Quality assurance:** Part of a management focused on providing confidence that quality requirements will be fulfilled.

38. **Quality control:** Part of a management focused on fulfilling quality requirements.

39. **Regulatory requirement:** Obligatory requirement specified by an authority mandated by legislative body.

40. **Release:** Permission to proceed to the next stage of a process or the next process.

41. **Requirement:** Need or expectation that is stated, generally implied or obligatory.

42. **Review:** Determination of the suitability, adequacy or effectiveness of an object to achieve established objectives.

43. **Risk:** Effect of uncertainty.

44. **Service:** Output of an organization with at least one activity necessarily performed between the organization and the customer.

45. **Statutory requirement:** Obligatory requirement specified by an authority mandated by a legislative body.

46. **Strategy:** Plan to achieve a long-term or overall objective.

47. **Success:** Achievement of an objective.

48. **Top management:** Person or group of people who directs and controls and organization at highest level.

49. **Traceability:** Ability to trace the history, application or location of an object.

50. **Validation:** Confirmation, through the provision of objective evidence, that the requirements for a specific intended use or application have been fulfilled.

51. **Verification:** Confirmation, through the provision of objective evidence, that specified requirements have been fulfilled.

52. **Vision:** Aspiration of what an organization would like to become as expressed by top management.

BIBLIOGRAPHY

1. https://antarisconsulting.wordpress.com/2018/04/20/iso-9004-overview-april-2018/

2. https://asq.org/quality-resources/iso-9000

3. https://en.wikipedia.org/wiki/Annex_SL

4. https://en.wikipedia.org/wiki/International_Organization_for_Standardization

5. https://en.wikipedia.org/wiki/ISO_19011

6. https://en.wikipedia.org/wiki/ISO_9000

7. https://insights.pecb.com/key-changes-iso-19011-2018/

8. https://www.iso.org/obp/ui/#iso:std:iso:9000:en

9. https://www.iso.org/sites/directives/2017/consolidated/index.xhtml#_idTextAnchor536

10. https://www.iso.org/standard/45481.html

11. https://www.qualitymag.com/articles/84855-quality-management-the-importance-of-iso
12. ISO 9000:2015 Standard

12. ISO 45001:2018 Standard

13. ISO 9001 Quality management system—requirements.

14. ISO 9004 Quality management—quality of an organization—guidance to achieve sustained success

15. ISO14001:2015 Standard

4

ISO 9000
Quality Management System
Requirements and Guidelines

OVERVIEW

ISO 9001 is the most important standard for quality managment system which has to be followed to achieve desired quality in construction. ISO 9000 standard is for normative reference and it is necessary to know these provisions in the context of ISO 9001. Similarly ISO 9004 gives guidance for sustained success. It is therefore necessary to understand interrelated provisions of ISO 9000 family of standares. For sustained success of the organization, it is important to fully appreciate specific requirements and guidelines of ISO 9000 family of standards.

ISO 9001:2015 contains specified requirements in various clauses which are descriptive but not fully prescriptive. Each organization has the flexibility to address the required subject clause in a manner which is most appropriate. The application of clauses will differ for an architectural/engineering company of a company of a contractor or a sub-contractor. With an effective and suitable approach, the required results can be achieved.

In the presentation of this chapter, consideration is given to specific requirements of the ISO 9001 standard and its application to the various organizations in the construction loop. In this chapter specific requirements of applicable clauses 4 to 10 of ISO 9001 along with corresponding provisions of ISO 9000 standard and ISO 9004 have been elaborated.

4.1 CLAUSE 4: CONTEXT OF THE ORGANIZATION

4.1.1 Requirement: Clause 4.1—Understanding the Organization and its Context

The ISO 9001:2015 standard requires that organization understands its external and internal issues (issues means both positive and negative factors or conditions for consideration) that are relevant to its purpose and its strategic direction and that affect its ability to achieve the intended result(s) of its quality management system. This can be achieved by determining internal and external issues which are relevant to QMS and monitor and review its related information.

External issues can be related to legal, technological, competitive, market, cultural, social and economic environments, whether international, national, regional or local.

Internal issues can be related to values, culture, knowledge and performance of the organization.

a. Guidelines: Clause 2.2.3 ISO 9000:2015

Understanding the context of an organization is one of the fundamental quality management concepts. Context of an organization is a process that determines factors influencing the organization's purpose, objectives and sustainability. The organization's purpose can be expressed through its vision, mission, policies and objectives.

b. Guidelines: Clause 3.2.2 ISO 9000:2015

Context of the organization is one of the terms related to organization concept class. The term is defined in the standard as "Combination of internal and external issues that can have an effect on an organization's approach to developing and achieving its objectives". In English, this concept is often referred to by other terms such as "business environment", "organizational environment" or "eco system of an organization". The concept of context of the organization is equally applicable to not for profit or public service organization as it is to those seeking profit. The organization's objectives can be related to its products, services, investments and behaviour towards its interested parties.

c. Guidelines: Clause 4.2 ISO 9004:2015

As the context of organization is ever-changing, to achieve sustained success top management should regularly monitor, analyse, evaluate and review the organization's context in order to identify all interested parties. Establishing a process for improvement, learning and innovation is important in order to support the organization's ability to respond to changes in the context of the organization.

d. Guidelines: Clause 5 ISO 9004:2018

When considering external and internal issues, the organization should take into account relevant information from the past, its current situation and its strategic direction. The organization should determine which external and internal issues could result in risks to its sustained success or opportunities to enhance its sustained success. Based on determination of these issues, top management should decide which of these risks and opportunities are to be addressed and initiate the establishment, implementation and maintenance of the necessary processes. The organization should consider how to establish, implement and maintain a process for monitoring, reviewing and evaluating external and internal issues, with consideration of any consequences to be acted on.

External issues are factors that exist outside of the organization that can affect the organization's ability to achieve sustained success, such as statutory and regulatory requirements, sector-specific requirements and agreements, competition, globalization, social, economic, political and cultural factors, innovations and advances in technology, natural environment.

Internal issues are factors that exist within the organization itself that can affect the organization's ability to achieve sustained success, such as size and complexity, activities and associated process, strategy, type of product and services, performance, resources, levels of competence and organizational knowledge, maturity and innovation.

e. Guidelines: Clause 6.2 ISO 9004:2018

Top management should ensure that the context of organization is considered when determining its mission, vision and values. Top management should review the mission, vision, values and culture at planned intervals and whenever the context of organization changes. This review should consider external and internal issues that can have effect on organization's ability to achieve sustained success.

4.1.2 Requirement: Clause 4.2 Understanding the Needs and Expectations of Interested Parties

Besides understanding internal and external issues, another key factor to be considered for the context of the organization is determining the needs and expectations of interested parties. The standard specifies requirements for the organization to determine the interested parties that are relevant to the quality management system and the requirements of the interested parties. The organizations need to monitor and review information about these interested parties and their relevant requirements.

a. Guidelines: Annex A (A.3) ISO 9001:2015

Sub-clause 4.2 specifies requirements for the organization to determine the interested parties that are relevant to the quality management system and the requirements of those interested parties.

However, 4.2 does not imply extension of quality management system requirements beyond the scope of ISO 9001:2015 standard. As stated in the scope, ISO 9001:2015 is applicable where an organization needs to demonstrate its ability to consistently provide products and services that meet customer and applicable statutory and regulatory requirements, and aims to enhance customer satisfaction.

There is no requirement for the organization to consider interested parties where it has decided that those interested parties are not relevant to quality management system. It is for the organization to decide if a particular requirement of a relevant interested party is relevant to its quality management system.

b. Guidelines: Clause 2.2.4 ISO 9000:2015

The concept of interested parties extends beyond a focus solely on the customer. It is important to consider all relevant interested parties.

Part of the process for understanding the context of organization is to identify its interested parties. The relevant interested parties are those that provide significant risk to organizational sustainability if their needs and expectations are not met. Organizations define what results are necessary to deliver to those relevant interested parties to reduce that risk. Organizations attract, capture and retain the support of the relevant interested parties they depend upon for their success.

c. Guidelines: Clause 3.2.3 ISO 9000:2015

Interested party is defined as person or organization that can affect, be affected by, or perceived itself to be affected by a decision or activity of the organization.

The examples of interested party are customers, owners, people in an organization, providers, bankers, regulators, unions, partners or society that can include competitors or opposing pressure groups.

d. Guidelines: Clause 4.1 ISO 9004:2015

The organization should go beyond the quality of its products and services and the needs and expectations of its customers. To achieve sustained success, the organization should focus on anticipating and meeting the needs and expectations of its interested parties, with the intent of enhancing their satisfaction and overall experience.

The organization should apply all of the quality management principles to achieve sustained success. Particular attention should be given to the principles of "customer focus" and "relationship management" to meet the different needs and expectations of interested parties.

The needs and expectations of individual interested parties can be different, aligned to, or in conflict with those of other interested parties, or can change quickly. The means by which the needs and expectations of interested parties are expressed and met can take wide variety of forms, for example, co-operation, negotiation, outsourcing, or by terminating an activity; consequently, the organization should give consideration to the interrelationships of its interested parties when addressing their needs and expectations.

The composition of interested parties can differ significantly over time and between organizations, industries, cultures and nations.

e. Guidelines: Clause 4.2 ISO 9004:2015

The quality of an organization is enhanced and sustained success can be achieved by consistently meeting needs and expectations of its interested parties over the long term. Short and medium term objectives should support this long term strategy. As the context of organization is ever-changing, to achieve sustained success top management should regularly monitor, analyse, evaluate and review the organization's context in order to identify all interested parties, determine their needs and expectations and their individual potential impacts on the organization's performance. Considering the needs and expectations of interested parties can enable the organization:

a. To achieve objectives effectively and efficiently
b. To eliminate the conflicting responsibilities and relationships
c. To harmonize and optimize practices
d. To create consistency
e. To improve communication
f. To facilitate training, learning and personal development
g. To facilitate focus on the most important characteristics of the organization.
h. To manage risks and opportunities to its brand or reputation
i. To acquire and share knowledge

f. Guideline: Clause 5.2 ISO 9004:2015

The organization should determine which interested parties are relevant. These interested parties can be both external and internal, including customers, and can impact the organization's ability to achieve sustained success. The organization should determine which interested parties:

    i. Are a risk to its sustained success if their relevant needs and expectations are not met.

    ii. Can provide opportunities to enhance its sustained success.

Once the relevant interested parties are determined, the organization should:

- identify their needs and expectations, determining the one that should be addressed.
- establish the necessary processes to fulfill the needs and expectations of the interested parties

The organization should consider how to establish ongoing relationships with the interested parties for benefits such as improved performance, common understanding of objectives and values, and enhanced stability.

g. Guidelines: Clause 6.2 ISO 9004:2015

When changes are made to any of the identity elements (i.e. mission, vision, values and culture) of organization, they should be communicated within the organization and to interested parties as appropriate.

4.2 REQUIREMENT: CLASE 4.3 DETERMINING THE SCOPE OF THE QUALITY MANAGEMENT SYSTEM

The organization shall determine the boundaries and applicability of the quality management system to establish its scope.

When determining this scope, the organization shall consider:

- The external and internal issues referred to in 4.1;
- The requirements of relevant interested parties referred to in 4.2;
- The products and services of the organization.

The organization shall apply all the requirements of ISO 9001:2015 if they are applicable within the determined scope of its quality management system. The scope of the organization's quality management system shall be available and be maintained as documented information. The scope shall state the types of products and services covered, and provide justification for any requirement of this international standard that the organization determines is not applicable to the scope of its quality management system.

Conformity to this International Standard may only be claimed if the requirements determined as not being applicable do not affect the organization's ability or responsibility to ensure the conformity of its products and services and the enhancement of customer satisfaction.

a. Guidelines: Annex A (A.5) ISO 9001:2015

This international standard does not refer to "exclusions" in relation to the applicability of its requirements to the organization's quality management system. However, an organization can review the applicability of requirements due to the size or complexity of the organization, the management model it adopts, the range of the organization's activities and the nature of risks and opportunities it encounters. The requirements for applicability are addressed in 4.3, which defines conditions under which an organization can decide that a requirement cannot be applied to any of the process within the scope of its quality management system.

The organization can only decide that a requirement is not applicable if its decision will not result in failure to achieve conformity of products and services.

b. Guidelines: Clause 3.5.3 ISO 9000:2015
The scope of a management system can include the whole of the organization, specific and identified functions of the organization, specific or identified sections of the organization, or one or more functions across a group of organizations.

4.3 REQUIREMENT: CLAUSE 4.4 QUALITY MANAGEMENT SYSTEM AND ITS PROCESSES

The organization shall establish, implement, maintain and continually improve a quality management system, including the processes needed and their interactions, in accordance with the requirements of ISO 9001:2015 standand.

The organization shall determine the processes needed for the quality management system and their application throughout the organization, and shall:
- Determine the inputs required and the outputs expected from these processes.
- Determine the sequence and interaction of these processes.
- Determine and apply the criteria and methods (including monitoring, measurements and related performance indicators) needed to ensure the effective operation and control of these processes.
- Determine the resources needed for these processes and ensure their availability;
- Assign the responsibilities and authorities for these processes.
- Address the risks and opportunities as determined in accordance with the requirements of ISO 9001:2015 standard.
- Evaluate these processes and implement any changes needed to ensure that these processes achieve their intended results.
- Improve the processes and the quality management system.

To the extent necessary, the organization shall:
- Maintain documented information to support the operation of its processes;
- Retain documented information to have confidence that the processes are being carried out as planned.

a. Guidelines: Introduction (0.3) ISO 9001:2015
The standard promotes adoption of a process approach when developing, implementing and improving the effectiveness of a QMS, to enhance customer satisfaction by meeting customer requirements. Specific requirements considered essential to the adoption of process approach are included in 4.4.

Understanding and managing interrelated processes as a system contributes to the organization's effectiveness and efficiency in achieving its intended results. This approach enables organization to control the interrelationships and interdependencies among the processes of the system, so that the overall performance of the organization can be enhanced.

The process approach involves the systematic definition and management of processes, and their interactions, so as to achieve the intended results in accordance with the quality policy and strategic direction of the organization. Management of the processes and the system as a whole can be achieved using the PDCA cycle with an overall focus on risk-based thinking aimed at taking advantage of opportunities and preventing undesirable results.

The application of the process approach in quality management system enables:

    i. Understanding and consistency in meeting requirements
    ii. The consideration of processes in terms of added value
    iii. The achievement of effective process performance
    iv. Improvement of processes based on evaluation of data and information

b. Guideline: Clause 2.3.4 ISO 9000:2015

Consistent and predictable results are achieved more effectively and efficiently when activities are understood and managed as interrelated processes that functions as a coherent system.

A QMS consists of interrelated processes. Understanding how results are produced by this system enables an organization to optimize the system and its performance.

Some potential key benefits of process approach are: Enhanced ability to focus on effort on key processes and opportunities for improvement, consistent and predictable outcomes through a system of aligned processes, optimized performance through effective process management, efficient use of resources and reduced cross-functional barriers, enabling the organization to provide confidence to interested parties related to its consistency, effectiveness and efficiency.

Possible actions include: Define objectives of the system and processes necessary to achieve them, establish authority; responsibility and accountability for managing processes; understand the organization's capabilities and determine resource constraints prior to action; determine process interdependencies and analyze the effect of modifications to individual processes on the system as a whole; manage processes and their interrelationships as a system to achieve the organization's quality objectives effectively and efficiently; ensure the necessary information is available to operate and improve the processes and to monitor; analyze and evaluate the performance of the overall system; manage risks which can affect outputs of the processes and overall outcome of the QMS.

c. Guidelines: Clause 3.5.3 and Clause 3.5.4 ISO 9000:2015

Management system is set of interrelated or interacting elements of an organization to establish policies and objectives and processes to achieve those objectives.

A management system can address a single discipline or several disciplines, e.g. quality management, financial management or environmental management.

The management system elements establish the organization's structure, roles and responsibilities, planning, operation, policies, practices, rules, beliefs, objectives and processes to achieve those objectives.

Quality management system is defined as part of management system with regard to quality.

d. Guidelines: Clause 3.4.1 ISO 9000:2015

Process is defined as set of interrelated or interacting activities that use inputs to deliver an intended result. Whether the "intended result" of a process is called output, product or service depends on the context of reference. Inputs to a process are generally the outputs of the other processes and outputs of a process are generally inputs to other process.

Processes in an organization are generally planned and carried out under controlled conditions to add value. A process where the conformity of resulting output cannot be readily or economically validated is frequently referred to as a "special process".

e. Guidelines: Clause 8 ISO 9004:2018
Organizations deliver value through activities connected within a network of processes. Processes often cross boundaries of functions within the organization. Consistent and predictable results are achieved more effectively and efficiently when the network of processes functions as a coherent system.

Processes are specific to an organization and vary depending on the type, size and level of maturity of the organization. The activities within each process should be determined and adapted to the size and distinctive features of the organization.

In order to achieve its objectives, the organization should ensure that all its processes are managed proactively, including externally provided processes, to ensure that they are effective and efficient. It is important to optimize the balance between the different purposes and specific objectives of the processes, in alignment with the organization's objectives.

This can be facilitated by adopting a "process approach" that includes establishing processes, interdependencies, constraints and shared resources.

The organization should determine the processes and their interactions necessary for providing outputs that meet the needs and expectations of interested parties, on an ongoing basis.

Processes and their interactions should be determined in accordance with the organization's policies, strategy, objectives and should deal with areas such as:
- Operations related to products and services
- Meeting the needs and expectations of interested parties
- The provision of resources
- Managerial activities, including monitoring, measuring, analysis, review, improvement, learning and innovation.

In determining its processes and their interactions, the organization should give consideration, as appropriate, to:
- The purpose of the process
- The objectives to be achieved and related performance indicators
- The outputs to be provided
- The needs and expectations of interested parties, and their changes
- Changes in operations, markets and technologies
- The impacts of the processes
- The inputs, resources and information needed, and their availability
- The activities that need to be implemented and methods that can be used
- Constraints for the process
- Risks and opportunities.

For each process, the organization should appoint a person or a team (often referred to as the "process owner"), depending on the nature of the process and the organization's culture, with defined responsibilities and authorities to determine, maintain, control and improve the process and its interaction with other processes it impacts and those that have impact on it. The organization should ensure that the responsibilities, authorities and roles of process owners are recognized throughout the organizations and that the people associated with the individual processes have the competences needed for the tasks and activities involved.

To manage its processes effectively and efficiently, the organization should:
- Manage the processes and their interactions, including externally provided processes, as a system to enhance alignment/linkage between the processes.
- Visualize the network of processes, their sequence and interactions in a graphic (e.g. process map, diagrams) in order to understand the roles of each process in the system and its effects on the performance of system.
- Determine criteria for the outputs of the processes, evaluate the capability and performance of processes by comparing the outputs with the criteria, and plan actions to improve the processes when they are not effectively achieving the performance expected by the system.
- Assess the risks and opportunities associated with the processes and implement any actions that are necessary in order to prevent, detect and mitigate undesired events, including risks such as:
  - Human factor (e.g. shortage of knowledge and skills, rule violations, human errors)
  - Inadequate capability, deteriorations and breakdowns of equipment
  - Design and development failure
  - Unplanned changes in incoming materials and services
  - Uncontrolled variations in the environment for the operation of processes
  - Unexpected changes in the needs and expectations of interested parties, including market demand
- Review the processes and their interrelationships on a regular basis and take suitable actions for control and improvement, to ensure they continue to be effective and support the sustained success of the organization.

Processes should operate together within a coherent management system. Some processes will relate to the overall management system and some will additionally relate to a specific managerial aspect, such as:
- The quality of products and services, including cost, quantity and delivery (e.g. ISO 9001).
- Health, safety, security (e.g. ISO 45001, ISO/IEC 27001).
- Environment, energy (e.g. ISO 14001, ISO 50001).
- Social responsibility, anti-bribery, compliance (e.g. ISO 26000, ISO 37001, ISO 19600)
- Business continuity, resilience (e.g. ISO 22301, ISO 22316).

To attain a higher level of performance, the processes and their interactions should be continually improved according to the organization's policies, strategy and objectives, including consideration of the need to develop or acquire new technologies, or to develop new products and services or their features, for added value.

The organization should motivate people to engage in improvement activities and propose opportunities for improvement in the processes and their interactions, the progress of action plans, and effects on the organization's policies, objectives and strategies. It should take any necessary corrective actions, or other appropriate actions, when gaps are identified between the planned and actual activities.

To maintain the level of performance attained, processes should be operated under controlled conditions, regardless of any planned or unplanned changes. The organization should determine what procedures (if any) are needed to manage the process, including the criteria for process outputs and operational conditions, to ensure conformity with the acceptance criteria.

When procedures are applied, in order to ensure that they are followed by people involved in the operation of the process, the organization should ensure that:

- A system is established to define the knowledge and skills needed for processes and evaluating the knowledge and skills of process operators.
- Risks in the procedures are identified, assessed and reduced by improving the procedures (e.g. making it difficult to make errors or not allowing progression to next process if any error occurs).
- Resources necessary for people to follow the procedures are made available.
- People have knowledge and skills needed for the following the procedures.
- People understand the impacts of not following the procedures (e.g. by using examples of experienced undesired events) and managers at appropriate levels take the actions that are necessary whenever a procedure is not followed.
- Consideration is given to learning, training, motivation and prevention of human error.

The organization should monitor its processes on a regular basis to detect deviations, and should identify and take appropriate actions when necessary without delay. Deviations are mainly caused by changes in equipment, method, material, measurement, environment and people for the operation of processes. The organization should determine check points and related performance indicators that will be effective and efficient in detecting deviations.

4.4 KEY REQUIREMENTS OF CLAUSE 4

This is a new clause that establishes the context of the QMS and how the business strategy supports this. The "context of the organization" is the clause that underpins the rest of the new standard. It gives an organization the opportunity to identify and understand the factors and parties in their environment that support the quality management system.

Firstly, the organization will need to determine external and internal issues that are relevant to its purpose, i.e. what are relevant issues, both inside and out, that have an impact on what the organization does, or that would affect its ability to achieve the intended outcome(s) of its management system.

It should be noted that the term "issue" covers not only the problems which would have been the subject of preventive action in the previous standards, but also important topics of management system to address, such as any market assurance and governance goals that the organization might set.

Secondly, an organization will also need to identify the "interested parties" that are relevant to their QMS. These groups could include shareholders, employees, customers, suppliers and even pressure groups and regulatory bodies. Each organization will identify their unique set of "interested parties" and over times these may change in line with the strategic direction of the organization.

Next the scope of the QMS must be determined. This could include the whole of the organization or specific identified functions. Any outsourced functions or processes will also need to be considered in the organization's scope if they are relevant to QMS. The final requirement of clause 4 is to establish, implement, maintain and continually improve the QMS in accordance with the requirements of the standard. This requires adoption of process approach and although every organization will be different. Documented information such as process diagrams or written procedures could be used to support this. ISO 9001:2015 contains numerous implicit documentation requirement.

In many construction projects like public buildings, hospitals, roads, etc. the main problems of quality in construction are various defects that become apparent at later stage when product is utilized by end user and rather than direct customer. To understand the quality of construction in detail, the final product has to be used for its design life.

The uniqueness of construction industry is that the final product cannot be tested completely. Therefore *in situ* quality in each process is more important and doing right at first time approach is more relevant at each stage of construction.

4.5 LEADERSHIP

4.5.1 Requirement: Clause 5.1.1 Leadership and Commitment—General

Top management shall demonstrate leadership and commitment with respect to the quality management system by:

- Taking accountability for the effectiveness of the quality management system;
- Ensuring that the quality policy and quality objectives are established for the quality management system and are compatible with the context and strategic direction of the organization;
- Ensuring the integration of the quality management system requirements into the organization's business processes;
- Promoting the use of the process approach and risk-based thinking;
- Ensuring that the resources needed for the quality management system are available;
- Communicating the importance of effective quality management and of conforming to the quality management system requirements;
- Ensuring that the quality management system achieves its intended results;
- Engaging, directing and supporting persons to contribute to the effectiveness of the quality management system;
- Promoting improvement.
- Supporting other relevant management roles to demonstrate their leadership as it applies to their areas of responsibility.

a. Guidelines: Clause 2.3.2 ISO 9000:2015

Leaders at all levels establish unity of purpose and direction and create conditions in which people are engaged in achieving the organization's quality objectives.

Creation of unity of purpose and the direction and engagement of people enable an organization to align its strategies, policies, processes and resources to achieve its objectives.

Some potential key benefits of Leadership are:

- Increased effectiveness and efficiency in meeting the organization's quality objectives.

- Better coordination of the organization's processes.
- Improved communication between levels and functions of the organization.
- Development and improvement of the capability of the organization and its people to deliver desired results.

Possible actions for leadership include:
- Communicate the organization's mission, vision, strategy, policies and processes throughout the organization.
- Create and sustain shared values, fairness and ethical models for behavior at all levels of the organization
- Establish a culture of trust and integrity
- Encourage an organization-wide commitment to quality
- Ensure that leaders at all levels are positive examples to people in the organization.
- Provide people with the required resources, training and authority to act with accountability
- Inspire, encourage and recognize the contribution of people.

b. Guidelines: Clause 3.1.1 ISO 9000:2015

Top management is one of the term related to person or people concept class and is defined as "Person or group of people who direct and control an organization at the highest level".

Top management has the power to delegate authority and provide resources within the organization. If the scope of management system covers only part of an organization, then top management refers to those who direct and control that part of the organization.

c. Guidelines: Clause 7 ISO 9004:2018

Top management, through its leadership, should:
- Promote the adoption of the mission, vision, values and culture in a way that is concise and easy to understand, to achieve unity of purpose.
- Create an internal environment in which people are engaged and committed to the achievement of organization's objectives.
- Encourage and support managers at appropriate levels to promote and maintain the unity of purpose and direction as established by the top management.

To achieve sustained success, top management should demonstrate leadership and commitment within the organization, by:
- Establishing the organization's identity
- Promoting a culture of trust and integrity
- Establishing and maintaining teamwork
- Providing people with the necessary resources, training and authority to act with accountability
- Promoting shared values, fairness and ethical behaviour so that these are sustained at all levels of organization.
- Establishing and maintaining an organizational structure to enhance competitiveness, where applicable

- Individually and collectively reinforcing the organization's values
- Communicating achieved successes externally and internally, as appropriate.
- Establishing a basis for effective communication with people in the organization, discussing issues that have general impact, including financial impact, where applicable.
- Supporting leadership development at all levels of the organization.

4.5.2 Requirement: Clause 5.1.2 Customer Focus

Top management shall demonstrate leadership and commitment with respect to customer focus by ensuring that:

- Customer and applicable statutory and regulatory requirements are determined, understood and consistently met.
- The risks and opportunities that can affect conformity of products and services and the ability to enhance customer satisfaction are determined and addressed.
- The focus on enhancing customer satisfaction is maintained.

a. Guidelines: Clause 3.2.3 ISO 9000:2015

Customer is an example of the interested party. Other interested parties can be owners, people in the organization, providers, bankers, regulators, unions, partners or society that can include competitors or opposing pressure groups.

b. Guidelines: Clause 3.2.4 ISO 9000:2015

Customer is one of the term related to "organization" concept class and defined as "Person or organization that could or does receive a product or a service that is intended for or required by this person or organization".

The examples of customer can be consumer, client, end-user, retailer, receiver of product or service from an internal process. A customer can be internal or external to the organization.

c. Guidelines: Clause 3.9 ISO 9000:2015

- "Customer" is one of the concept class of ISO 9000:2015 defining terms related to customer: Feedback, customer satisfaction, complaint, customer service, customer satisfaction code of conduct and dispute.
- Feedback is opinions, comments and expressions of interest in a product, a service or a complaints-handling process.
- Customer satisfaction is customer's perception of the degree to which the customer's expectations have been fulfilled. It can be that customer's expectation is not known to the organization, or even to the customer in question, until the product or service is delivered. It can be necessary for achieving high customer satisfaction to fulfill an expectation of a customer even if it is neither stated nor generally implied or obligatory. Complaints are common indicators of low customer satisfaction but their absence does not necessarily imply high customer satisfaction. Even when customer requirements have been agreed with the customer and fulfilled, this does not necessarily ensure high customer satisfaction.
- Complaint is expression of dissatisfaction made to an organization, related to its product or service or the complaints handling process itself, where a response or resolution is implicitly or explicitly expected.

- Customer service is interaction of the organization with the customer throughout the life cycle of product or service.
- Customer satisfaction code of conduct is promises, made to customers by an organization concerning its behaviour, that are aimed at enhanced customer satisfaction and related provisions. Related provisions can include objectives, conditions, limitations, contract information and complaints handling procedures.
- Dispute is disagreement, arising from complaint, submitted to a DRP-provider (dispute resolution process provider). Some organizations allow their customers to express their dissatisfaction to a DRP provider in the first instance. In this situation, the expression of dissatisfaction becomes a complaint when sent to organization for response, and becomes a dispute if not resolved by the organization without DRP provider intervention. Many organizations prefer their customers to first express any dissatisfaction to the organization before utilizing dispute resolution external to the organization.

4.5.3 Requirement: Clause 5.2 Policy

Top management shall establish, implement and maintain a quality policy that: (a) is appropriate to the purpose and context of the organization and supports its strategic direction; (b) provides a framework for setting quality objectives; (c) includes a commitment to satisfy applicable requirements; (d) includes a commitment to continual improvement of the quality management system.

The quality policy shall be: (a) available and maintained as documented information; (b) communicated, understood and applied within the organization; (c) be available to relevant interested parties, as appropriate.

a. Guidelines: Clause 3.5.8 ISO 9000:2015

Policy is defined as intentions and directions of an organization as formally expressed by its top management.

b. Guidelines: Clause 7.2 ISO 9004:2018

Top management should set out the organization's intentions and direction in the form of the organization's policy, to address aspects such as compliance, quality, environment, energy, employment, occupational health and safety, quality of work life, innovation, security, privacy, data protection and customer experience. Policy statements should include commitments to satisfy the needs and expectations of interested parties and to promote improvement. The policy decisions should be reviewed for continued suitability. Any changes to external and internal issues, as well as any new risks and opportunities should be addressed.

4.5.4 Requirement: Clause 5.3 Organizational Roles, Responsibilities and Authorities

Top management shall ensure that the responsibilities and authorities for relevant roles are assigned, communicated and understood within the organization.

Top management shall assign the responsibility and authority for:

- Ensuring that the quality management system conforms to the requirements of this international standard;
- Ensuring that the processes are delivering their intended outputs;
- Reporting on the performance of the quality management system and on opportunities for improvement, in particular to top management;
- Ensuring the promotion of customer focus throughout the organization;
- Ensuring that the integrity of the quality management system is maintained when changes to the quality management system are planned and implemented.

a. Guidelines: Clause 8.3 ISO 9004:2018

For each process, the organization should appoint a person or a team (often referred to as the "process owner"), depending on the nature of the process and the organization's culture, with defined responsibilities and authorities to determine, maintain, control and improve the process and its interaction with other processes it impacts and those that have impact on it. The organization should ensure that the responsibilities, authorities and roles of process owners are recognized throughout the organization and that the people associated with the individual processes have the competence needed for the tasks and activities involved.

4.5.5 Key Requirements of Clause 5

This clause places requirements on "top management" which is the person or group of people who directs and controls the organization at highest level. It is no longer the responsibility of an individual or to have "Management Representative" who is responsible for QMS. There is an increased emphasis on people "owning" the QMS rather than one individual. The purpose of these requirements is to demonstrate leadership and commitment by leading from the top.

Top management now have greater involvement in the management system and must ensure that the requirements of it are integrated into the organization's processes and that the policy and objectives are compatible with the strategic direction of the organization. The quality policy should be a living document, at the heart of the organization. To ensure this, top management are accountable and have a responsibility to ensure the QMS is made available, communicated, maintained and understood by all parties.

There is also greater focus on top management to enhance customer satisfaction by identifying and addressing risks and opportunities that could affect this. Top management need to demonstrate consistent customer focus by showing how they meet customer requirements, regulatory and statutory requirements, and also how the organization maintains enhanced customer satisfaction.

In the same context, they need to have a grasp of the organizations internal strengths and weaknesses and how these could have an impact to deliver products or services. This will strengthen the concept of business process management. In addition, top management need to demonstrate an understanding of the key risks associated with each process and the approach taken to manage, reduce or transfer the risk.

Finally, the clause places requirements on the top management to assign QMS relevant responsibilities and authorities, but must remain accountable for the effectiveness of the QMS.

4.6 REQUIREMENT CLAUSE 6.1: PLANNING ACTIONS TO ADDRESS RISKS AND OPPORTUNITIES

When planning for the quality management system, the organization shall consider the internal and external issues and the requirements of interested parties and determine the risks and opportunities that need to be addressed to: (a) Give assurance that the quality management system can achieve its intended result(s); (b) enhance desirable effects; (c) prevent, or reduce, undesired effects; (d) achieve improvement.

The organization shall plan: (a) Actions to address these risks and opportunities; (b) how to: (1) Integrate and implement the actions into its quality management system processes; (2) evaluate the effectiveness of these actions.

Actions taken to address risks and opportunities shall be proportionate to the potential impact on the conformity of products and services. Options to address risks can include avoiding risk, taking risk in order to pursue an opportunity, eliminating the risk source, changing the likelihood or consequences, sharing the risk, or retaining risk by informed decision. Opportunities can lead to the adoption of new practices, launching new products, opening new markets, addressing new customers, building partnerships, using new technology and other desirable and viable possibilities to address the organization's or its customers' needs.

a. Guidelines: Introduction (0.3.3) ISO 9001:2015

Risk-based thinking is essential for achieving an effective quality management system. The concept of risk-based thinking has been implicit in previous editions of ISO 9001 standard including, for example, carrying out preventive action to eliminate potential nonconformities, analyzing any nonconformities that do occur, and taking action to prevent recurrence that is appropriate for the effects of nonconformity.

To conform to the requirements of ISO 9001 standard, an organization needs to plan and implement actions to address risks and opportunities. Addressing both risks and opportunities establishes a basis for increasing the effectiveness of the quality management system, achieving improved results and preventing negative effects.

Opportunities can arise as a result of a situation favourable to achieving an intended result, for example, a set of circumstances that allow the organization to attract customers, develop new products and services, reduce waste or improve productivity. Actions to address opportunities can also include consideration of associated risks. Risk is the effect of uncertainty and any such uncertainty can have positive or negative effects. A positive deviation arising from a risk can provide an opportunity, but not all positive effects of risk result in opportunity.

b. Guidelines: Annex A (A.4) ISO 9001:2015

The concept of risk-based thinking has been implicit in previous editions of ISO 9001, e.g. through requirements for planning, review and improvement. This international standard specifies requirements for the organization to understand its context and determine risks as a basis for planning. This represents the application of risk-based thinking to planning and implementing quality management system processes and will assist in determining the extent of documented information.

One of the key purposes of a quality management system is to act as a preventive tool. Consequently, ISO 9001: 2015 standard does not have a separate clause or sub-clause on preventive action. The concept of preventive action is expressed through the use of risk-based thinking in formulating quality management system requirements.

The risk-based thinking applied in ISO 9001:2015 standard has enabled some reduction in prescriptive requirements and their replacement by performance-based requirements. There is greater flexibility than in ISO 9001:2008 in the requirements for processes, documented information and organizational responsibilities.

Although ISO 9001:2015 Clause 6.1—actions to address risks and opportunities, specifies that the organization shall plan actions to address risks, there is no requirement for a formal method for risk management or a documented risk management process. Organizations can decide whether or not to develop a more extensive risk management methodology that is required by ISO 9001:2015 standard, e.g. through the application of other guidance or standards.

Not all the processes of quality management system represent the same level of risk in terms of the organization's ability to meet its objectives, and the effects of uncertainty are not same for all organizations. The organization is responsible for risk based thinking and the actions it takes to address risk, including whether or not to retain documented information as evidence of its determination of risks.

c. Guidelines: Clause 3.7.9 ISO 9000:2015

"Risk" is a quality management term related to result concept class. It is defined as effect of uncertainty. An effect is deviation from the expected-positive or negative. Uncertainty is the state, even partial, of deficiency of information related to, understanding or knowledge of, an event, its consequence or likelihood. Risk is often characterized by reference to potential events and consequences, or a combination of these. Risk is often expressed in terms of a combination of the consequences of an event (including changes in circumstances) and the associated likelihood of occurrence. The word "risk" is sometimes used when there is possibility of only negative consequences.

d. Guidelines: Clause 8.4.1 ISO 9004:2018

The organization should assess the risks and opportunities, associated with the processes and implement any actions that are necessary in order to prevent, detect and mitigate undesired events, including risks such as:

a. Human factors (e.g. shortage of knowledge and skills, rule violations, human errors).
b. Inadequate capability, deteriorations and breakdowns of equipment.
c. Design and development failures.
d. Unplanned changes in incoming materials and services.
e. Uncontrolled variation in the environment for the operation of processes.
f. Unexpected changes in the needs and expectations of interested parties, including market demand.

e. Guidelines: Clause 11.4.3 ISO 9004:2018

The organization should evaluate the risks and opportunities related to its plans for innovation activities. It should give consideration to the potential impact on the managing of changes and prepare action plans to mitigate those risks (including contingencies plans), where necessary. The timing for introduction of an innovation should be aligned with the evaluation of the risk associated with undertaking that innovation.

f. Guidelines

One of the key changes in the 2015 revision of ISO 9001 is to establish a systematic approach to considering risk, rather than treating "prevention" as a separate component of a quality management system. Risk is inherent in all aspects of a quality management system. In ISO 9001:2015 risk-based thinking needs to be considered from the beginning and throughout the system, making preventive action inherent to planning, operation, analysis and evaluation activities. Risk-based thinking ensures these risks are identified, considered and controlled throughout the design and use of the quality management system.

ISO 9001:2015 uses risk-based thinking in the following way: (1) In Introduction the concept of risk-based thinking is explained, (2) the clause 4 requires organization to address the risks and opportunities associated with its QMS processes, (3) The clause 5 requires top management to promote awareness of risk-based thinking and determine and address risks and opportunities that can affect product/service conformity, (4) the clause 6 requires organizations to identify risks and opportunities related to QMS performance and take appropriate actions to address them, (5) the clause 7 requires organizations to determine and provide necessary resources (risk is implicit whenever "suitable" or "appropriate" is mentioned), (6) the clause 8 requires organizations to manage its operational processes (risk is implicit whenever "suitable" or "appropriate" is mentioned), (7) the clause 9 requires organization to monitor, measure, analyse and evaluate effectiveness of actions taken to address the risks and opportunities, (8) the clause 10 requires organizations to correct, prevent or reduce undesired effects and improve the QMS and update risks and opportunities.

Comprehensively risk-based thinking: (1) is not new, (2) is something you do already, (3) is continuous, (4) ensures greater knowledge of risks and improves preparedness, (5) increases the probability of reaching objectives, (6) reduces the probability of negative results, (7) makes prevention a habit.

Other useful documents: (1) ISO 31000:2009 risk management—principles and guidelines, (2) ISO 31010:2010 risk management—risk assessment techniques, (3) PD ISO/TR 31004:2013 risk management—guidance for the implementation of ISO 31000.

4.6.1 Requirement: Clause 6.2: Quality Objectives and Planning to Achieve them

The organization shall establish quality objectives at relevant functions, levels and processes needed for the quality management system. The quality objectives shall—(a) be consistent with the quality policy, (b) be measurable, (c) take into account applicable requirements, (d) be relevant to conformity of products and services and to enhancement of customer satisfaction,

(e) be monitored, (f) be communicated, (g) be updated as appropriate. The organization shall maintain documented information on the quality objectives.

When planning how to achieve its quality objectives, the organization shall determine: (a) what will be done; (b) what resources will be required; (c) who will be responsible; (d) when it will be completed; (e) how the results will be evaluated.

a. Guidelines: Clause 3.7.1 and Clause 3.7.2 ISO 9000:2015

Objective and quality objective both are quality management terms related to result concept class. Objective is defined as result to be achieved and quality objective is defined as objectives related to quality. An objective can be strategic, tactical or operational. Objectives can relate to different disciplines (such as financial, health and safety, and environmental objectives) and can apply at different levels (such as strategic, organization-wide, project, product and process). An objective can be expressed in other ways, e.g. as an intended outcome, a purpose, an operational criterion, as a quality objective or by use of other words with similar meaning (e.g. aim, goal, or target).

In the context of quality management systems quality objectives are set by the organization, consistent with the quality policy, to achieve specific results. Quality objectives are generally based on the organization's quality policy. Quality objectives are generally specified for relevant functions, levels and processes in the organization.

b. Guidelines: Clause 7.3 ISO 9004:2018

Top management should demonstrate leadership in the organization by defining and maintaining the organization's objectives based on its policy and strategy, as well as by deploying the objectives at relevant functions, levels and processes.

Objectives should be defined for the short and long term and should be clearly understandable. Objectives should be quantified where possible. When determining the objectives, top management should consider:

- To what extent the organization is aiming to be recognized by interested parties as:
  - A leader with respect to competitive factors emphasizing the organization's capability.
  - Having a positive impact on economic, environmental and social conditions around it.
- The degree of the organization's and its peoples engagement in society beyond immediate business related topics (e.g. in national and international organizations, such as public administration, associations and standardization bodies)

When deploying the objectives, top management should encourage discussions for alignment between different functions and levels of the organization.

4.6.2 Requirement: Clause 6.3 Planning of Changes

When the organization determines the need for changes to the quality management system, the changes shall be carried out in a planned manner. The organization shall consider: (a) The purpose of the changes and their potential consequences, (b) the integrity of the quality management system, (c) the availability of resources, (d) the allocation or reallocation of responsibilities and authorities.

a. Guidelines: Clause 2.4.1 ISO 9000:2015

An organization's QMS model recognizes that not all systems, processes and activities can be predetermined; therefore it needs to be flexible and adaptable within the complexities of the organizational context.

A QMS is a dynamic system that evolves over time through periods of improvement. Every organization has quality management activities, whether they have been formally planned or not. It is necessary to determine activities which already exist in the organization and their suitability regarding the context of organization. ISO 9000 along with ISO 9004 and ISO 9001, can be used to assist the organization to develop a cohesive system.

A QMS does not need to be complicated; rather it needs to accurately reflect the needs of the organization. In developing the QMS, the fundamental concepts and principles given in ISO 9000:2015 can provide valuable guidance.

QMS planning is not a singular event, rather it is an ongoing process. Plans evolve as the organization learns and circumstances change. A plan takes into account all quality activities of the organization and ensures that it covers all guidance of ISO 9000:2015 and requirements of ISO 9001:2015.

b. Guidelines: Clause 11.1 ISO 9004:2018

The organization will experience constant change in its external and internal issues and in the needs and expectations of its interested parties. Improvement, learning and innovation support the organization's ability to respond to these changes in a manner that enables it to fulfill its mission and vision, as well as supporting its achievement of sustained success.

4.6.3 Key Requirements of Clause 6

Planning has always been a familiar element of ISO 9001, but now there is increased focus on ensuring that it is considered with Clause 4.1 and Clause 4.2 interested parties.

The first part of the clause concerns risk assessment whilst the second part is concerned with risk treatment. When determining actions to identify risks and opportunities these need to be proportionate to the potential impact they may have on conformity of products and services. Opportunities could, for example, include new product launches, geographical expansion, new partnerships, or new technologies.

The organization will need to plan actions to address both risks and opportunities, how to integrate and implement the actions into its management system processes and evaluate the effectiveness of these actions. Actions must be monitored, managed and communicated across the organization.

By considering risk throughout the system and all processes the likelihood of achieving stated objectives is improved, output is more consistent and customers can be confident that they will receive the expected product or service. Successful companies intuitively incorporate risk-based thinking. Risk management does not require the establishment of new department as it is an integrated function.

Another key element of this clause is the need to establish measurable quality objectives. Quality objectives now need to be consistent with the quality policy, relevant to the conformity of products and services as well as enhancing customer satisfaction.

The last part the clause considers planning of changes which must be done in a planned and systematic manner. There is a need to identify the potential consequences of changes, determine who is involved, when changes are to take place, what resource needs to be allocated.

4.7 CLAUSE 7 SUPPORT

4.7.1 Requirement: Clause 7.1 Resources—General

The organization shall determine and provide the resources needed for the establishment, implementation, maintenance and continual improvement of the quality management system. The organization shall consider: (a) the capabilities of, and constraints on, existing internal resources; (b) what needs to be obtained from external providers.

a. Guidelines: Clauses 3.2.5 and 3.2.6 ISO 9000:2015

Provider is defined as organization that provides a product or a service. A provider can be internal or external to the organization. External provider is provider that is not part of the organization, e.g. producer, distributor, retailer or vendor of a product or a service. In a contractual situation, a provider is sometimes called "Contractor".

b. Guidelines: Clause 9 ISO 9004:2018

Resources support the operation of all processes in an organization and are critical for ensuring effective and efficient performance and its sustained success. The organization should determine and manage the resources needed for the achievement of its objectives, taking into account the associated risks and opportunities and their potential effects. Examples of key resources include: (a) Financial resources; (b) People; (c) Organizational knowledge; (d) Technology; (e) Infrastructure, such as equipment, facilities, energy and utilities; (f) The environment for the organization's processes; (g) The materials needed for the provision of products and services; (h) Information; (i) Resources provided externally, including subsidiaries, partnerships and alliances; (j) Natural resources.

The organization should implement sufficient control over its processes to achieve efficient and effective use of its resources. Depending on the nature and complexity of the organization, some of the resources will have different impacts on the sustained success of the organization.

When considering future activities, the organization should take into account the accessibility and suitability of resources, including externally provided resources. The organization should frequently evaluate its existing use of resources to determine opportunities for improving their use, optimizing processes, and implementing new technologies to reduce risk.

4.7.2 Requirement: Clause 7.1.2 People

The organization shall determine and provide the persons necessary for the effective implementation of its quality management system and for the operation and control of its processes.

a. Guidelines: Clause 2.2.5.2 ISO 9000:2015

People are essential resources within the organization. The performance of the organization is dependent upon how people behave within the system in which they work. Within an organization, people become engaged and aligned through a common understanding of the quality policy and the organization's desired results.

b. Guidelines: Clause 2.3.3 ISO 9000:2015

Competent, empowered and engaged people at all levels throughout the organization are essential to enhance the organization's capability to create and deliver value.

In order to manage an organization effectively and efficiently, it is important to respect and involve all people at all levels. Recognition, empowerment and enhancement of competence facilitate the engagement of people in achieving the organization's quality objectives.

Some potential key benefits are: (a) Improved understanding of the organization's quality objectives by people in the organization and increased motivation to achieve them; (b) enhanced involvement of people in improvement activities; (c) enhanced personal development, initiatives and creativity; (d) enhanced people satisfaction; (e) enhanced trust and collaboration throughout the organization; (f) increased attention to shared values and culture throughout the organization.

Possible actions include: (a) Communicate with people to promote understanding of the importance of their individual contribution; (b) promote collaboration throughout the organization; (c) facilitate open discussion and sharing of knowledge and experience; (d) empower people to determine constraints to performance and to take initiatives without fear; (e) recognize and acknowledge people's contribution, learning and improvement; (f) enable self-evaluation of performance against personal objectives; (g) conduct surveys to assess people's satisfaction, communicate the results and take appropriate actions.

c. Guidelines: Clause 9.2 ISO 9004:2018

Competent, engaged, empowered and motivated people are a key resource. The organization should develop and implement processes to attract and retain people who have the current or potential competences and availability to contribute fully to the organization. The managing of people should be performed through a planned, transparent, ethical and socially responsible approach at all levels throughout the organization.

Engagement of people enhances the organization's ability to create and deliver value for interested parties. The organization should establish and maintain processes for engagement of its people. Managers at all levels should encourage people to be involved in improving performance and meeting the organization's objectives.

To enhance the engagement of its people, the organization should consider activities such as: (a) Developing a process to share knowledge; (b) making use of its people's competence; (c) establishing a skill qualification system and career planning to promote personal development; (d) continually reviewing their level of satisfaction, relevant needs and expectations; (e) providing opportunities for mentoring and coaching; (f) promoting team improvement activities.

Empowered and motivated people at all levels throughout the organization are essential to enhance the organization's ability to create and deliver value. Empowerment enhances the motivation of people to take responsibility for their work and its results. This can be achieved by providing people with the necessary information, authority and freedom to make decisions related to their own work. Managers at all levels should motivate people to understand the significance and importance of their responsibilities and activities in relation to creating value for interested parties. To enhance the empowerment and motivation of people, managers at all levels should:

- define clear objectives (aligned with the organization's objectives), delegate authority and responsibility, and create a work environment in which people control their own work and decision making;
- introduce an appropriate recognition system, based on the evaluation of people's accomplishments (both individually and in teams)
- provide incentives for people to act with initiative (both individually and in teams), as well as recognizing good performance, rewarding results and celebrating the achievement of objectives.

4.7.3 Requirement: Clause 7.1.3 Infrastructure

The organization shall determine, provide and maintain the infrastructure necessary for the operation of its processes and to achieve conformity of products and services. Infrastructure can include: (a) Buildings and associated utilities; (b) equipment, including hardware and software; (c) transportation resources; (d) information and communication technology.

a. Guidelines: Clause 3.5.2 ISO 9000:2015

Infrastructure is a quality management term related to system concept class and defined as, <organization> system of facilities, equipment and services needed for the operation of an organization.

b. Guidelines: Clause 9.5.2 ISO 9004:2018

In managing its infrastructure, the organization should give appropriate consideration to factors such as (a) Dependability (including consideration of availability, reliability, maintainability and maintenance support, as applicable, including safety and security); (b) infrastructure elements needed for the provision of processes, products and services; (c) the efficiency, capacity and investment required; d) the impact of infrastructure.

4.7.4 Requirement: Clause 7.1.4 Environment for the Operation of Processes

The organization shall determine, provide and maintain the environment necessary for the operation of its processes and to achieve conformity of products and services. A suitable environment can be a combination of human and physical factors, such as (a) social (e.g. non-discriminatory, calm, non-confrontational); (b) psychological (e.g. stress-reducing, burnout prevention, emotionally protective); (c) physical (e.g. temperature, heat, humidity, light, airflow, hygiene, noise). These factors can differ substantially depending on the products and services provided.

a. Guidelines: Clause 3.5.5 ISO 9000:2015

Set of conditions under which work is performed. Conditions can include physical, social, psychological and environmental factors (such as temperature, lighting, recognitions schemes, occupational stress, ergonomics and atmospheric composition)

b. Guidelines: Clause 9.5.3 ISO 9004:2018

In determining a suitable work environment, the organization should give appropriate consideration to factors (or a combination of factors) such as (a) physical characteristics such as heat, humidity, light, airflow, hygiene, cleanliness and noise; (b) ergonomically designed work stations and equipment; (c) psychological aspects; (d) encouraging personal growth; (e) creative work methods and opportunities for greater involvement, to realize the potential of people in the organization; (f) health and safety rules and guidance, as well as the use of protective equipment; (g) workplace location; (h) facilities for people in the organization; (i) optimization of resources.

The organization's work environment should encourage productivity, creativity and well-being for the people working in or visiting its premises (e.g. customers, external providers, partners). In addition, depending on its nature, the organization should verify that its work environment complies with applicable requirements and addresses applicable standards (such as those for environmental and occupational health and safety management).

4.7.5 Requirement: Clause 7.1.5 Monitoring and Measuring Resources

The organization shall determine and provide the resources needed to ensure valid and reliable results when monitoring or measuring is used to verify the conformity of products and services to requirements. The organization shall ensure that the resources provided are: (a) Suitable for the specific type of monitoring and measurement activities being undertaken; (b) Maintained to ensure their continuing fitness for their purpose. The organization shall retain appropriate documented information as evidence of fitness for purpose of the monitoring and measurement resources.

When measurement traceability is a requirement, or is considered by the organization to be an essential part of providing confidence in the validity of measurement results, measuring equipment shall be: (a) Calibrated or verified, or both, at specified intervals, or prior to use, against measurement standards traceable to international or national measurement standards; when no such standards exist, the basis used for calibration or verification shall be retained as documented information; (b) identified in order to determine their status; (c) safeguarded from adjustments, damage or deterioration that would invalidate the calibration status and subsequent measurement results. The organization shall determine if the validity of previous measurement results has been adversely affected when measuring equipment is found to be unfit for its intended purpose, and shall take appropriate action as necessary.

a. Guidelines: Clause 3.11 ISO 9000:2015

Monitoring is defined as determining the status of a system, a process, a product, a service or an activity; for determination of the status there can be a need to check, supervise or critically observe. Monitoring is generally a determination of the status of an object, carried out at different stages or at different times.

Measurement is defined as process to determine a value. The value is generally the value of a quantity. Measurement process is defined as set of operations to determine the value of a quantity. Measuring equipment is defined as measuring instrument, software, measurement standard, reference material or auxiliary apparatus or combination thereof necessary to realize a measurement process.

4.7.6 Requirement: Clause 7.1.6 Organizational Knowledge

The organization shall determine the knowledge necessary for the operation of its processes and to achieve conformity of products and services.

This knowledge shall be maintained and be made available to the extent necessary.

When addressing changing needs and trends, the organization shall consider its current knowledge and determine how to acquire or access any necessary additional knowledge and required updates. Organizational knowledge is knowledge specific to the organization; it is generally gained by experience. It is information that is used and shared to achieve the organization's objectives. Organizational knowledge can be based on: (a) internal sources (e.g. intellectual property; knowledge gained from experience; lessons learned from failures and successful projects; capturing and sharing undocumented knowledge and experience; the results of improvements in processes, products and services); (b) external sources (e.g. standards; academia; conferences; gathering knowledge from customers or external providers).

a. Guidelines: Annex A (A.7) ISO 9001:2015

ISO 9001:2015 addresses the need to determine and manage the knowledge maintained by the organization, to ensure the operation of its processes and that it can achieve conformity of products and services. Requirements regarding organizational knowledge were introduced for the purpose of: (a) Safeguarding the organizational knowledge from loss of knowledge, e.g. through staff turnover, failure to capture and share information; (b) encouraging organization to acquire knowledge, e.g. Learning from experience, mentoring, benchmarking.

b. Guidelines: Clause 9.3 ISO 9004:2018

Organizational knowledge can be based on external or internal resources. Top management should: (a) Recognize knowledge as an intellectual asset and manage it as an essential element of the organization's sustained success; (b) consider the knowledge required to support the short- and long-term needs of the organization, including succession planning; (c) assess how the organization's knowledge is identified, captured, analyzed, retrieved, maintained and protected.

When defining how to determine, maintain and protect knowledge, the organization should develop processes to address: (a) Lessons learned from failures and successful projects; (b) explicit and tacit knowledge that exists within the organization, including the knowledge, insights and experience of its people; (c) determining the need to acquire knowledge from interested parties as part of the organization's strategy; (d) conforming the effective distribution and understanding of information, throughout the life cycle(s) of the organization's products and services; (e) managing documented information and its use; (f) managing intellectual property.

c. Guidelines: Clause 11.3 ISO 9004:2018

A learning approach should be adopted by the organization as a whole, as well as at a level that integrates the capabilities of individuals with those of the organization. Learning as an organization involves consideration of:

- Collected information relating to various external and internal issues and interested parties, including success stories and failures;
- Insight through in-depth analysis of the information collected.

Learning that integrates the capabilities of individuals with those of the organization is achieved by combining the knowledge, thinking patterns and behaviour patterns of people with the values of the organization. Knowledge can be explicit or tacit. It can originate from inside or outside the organization. It should be managed and maintained as an asset of the organization. The organization should monitor its organizational knowledge and determine the need to acquire, or more effectively share, knowledge throughout the organization.

In order to foster learning organization, the following factors should be considered: (a) The organization's culture, aligned with its mission, vision and values; (b) top management supporting initiatives in learning, by demonstrating its leadership and through its behaviour; (c) stimulation of networking, connectivity, interactivity and sharing of knowledge both inside and outside of the organization; (d) maintaining system for learning and sharing of knowledge; (e) recognizing, supporting and rewarding the improvement of people's competence, through processes for learning and sharing of knowledge; (f) appreciating creativity and supporting diversity of the opinions of the different people in the organization. Rapid access to, and use of, organizational knowledge can enhance the organization's ability to manage and maintain its sustained success.

4.7.7 Requirement Clause 7.2 Competence

The organization shall: (a) Determine the necessary competence of person(s) doing work under its control that affects the performance and effectiveness of the quality management system; (b) ensure that these persons are competent on the basis of appropriate education, training, or experience; (c) where applicable, take actions to acquire the necessary competence, and evaluate the effectiveness of the actions taken; (d) retain appropriate documented information as evidence of competence. Applicable actions can include, for example, the provision of training to, the mentoring of, or the reassignment of currently employed persons; or the hiring or contracting of competent persons.

a. Guidelines: Clause 2.2.5.3 ISO 9000:2015

A QMS is most effective when all employees understand and apply skills, training, education and experience needed to perform their roles and responsibilities. It is the responsibility of top management to provide opportunities for people to develop theses necessary competencies.

b. Guidelines: Clause 9.2.4 ISO 9004:2018

A process should be established and maintained to assist the organization in determining, developing, evaluating and improving the competence of people at all levels. The process should follow steps such as (a) determining and analyzing the personal competencies needed by the organization in accordance with its identity (mission, vision, values and culture), strategy, policies and objectives; (b) determine the current competencies at group level and at individual level, as well as gaps available and what is currently needed, or could be needed in the future; (c) implementing actions to improve and acquire competence, as required. (d) improving and maintaining the competence that has been acquired; (e) reviewing and evaluating the effectiveness of actions taken to confirm that the necessary competence has been acquired.

4.7.8 Requirement: Clause 7.3 Awareness

The organization shall ensure that persons doing work under the organization's control are aware of: (a) The quality policy; (b) relevant quality objectives; (c) their contribution to the effectiveness of the quality management system, including the benefits of improved performance; (d) the implications of not conforming with the quality management system requirements.

a. Guidelines: Clause 2.2.5.4 ISO 9000:2015

Awareness is attained when people understand their responsibilities and how their actions contribute to the achievement of the organization's objectives.

4.7.9 Requirement: Clause 7.4 Communication

The organization shall determine the internal and external communications relevant to the quality management system, including: (a) on what it will communicate; (b) when to communicate; (c) with whom to communicate; (d) how to communicate; (e) who communicates.

a. Guidelines: Clause 2.2.5.5 ISO 9000:2015

Planned and effective internal (i.e. throughout the organization) and external (i.e. with relevant interested parties) communication enhances people's engagement and increased understanding of: (a) the context of the organization; (b) the needs and expectations of customers and other relevant interested parties; (c) the QMS.

b. Guidelines: Clause 7.4 ISO 9004:2018

The effective communication of policies and strategy with relevant objectives, is essential to support the sustained success of the organization. Such communication should be meaningful, timely and continual. Communication should include a feedback mechanism and incorporate provisions to proactively address changes in the organization's context. The organization's communication process should operate both vertically and horizontally and should be tailored to the differing needs of the recipients. For example, the same information can be conveyed in one way to people within the organization and in different way to interested parties.

4.7.10 Requirement: Clause 7.5 Documented Information

The organization's quality management system shall include: (a) documented information required by this international standard; (b) documented information determined by the organization as being necessary for the effectiveness of the quality management system. The extent of documented information for a quality management system can differ from one organization to another due to: The size of organization and its type of activities, processes, products and services; The complexity of processes and their interactions; The competence of persons.

When creating and updating documented information, the organization shall ensure appropriate: (a) Identification and description (e.g. a title, date, author, or reference number); (b) format (e.g. language, software version, graphics) and media (e.g. paper, electronic); (c) review and approval for suitability and adequacy.

Documented information required by the quality management system and by this international standard shall be controlled to ensure: (a) It is available and suitable for use, where and when it is needed; (b) it is adequately protected (e.g. from loss of confidentiality, improper use, or loss of integrity).

For the control of documented information, the organization shall address the following activities, as applicable: (a) distribution, access, retrieval and use; (b) storage and preservation, including preservation of legibility; (c) control of changes (e.g. version control); (d) retention and disposition.

Documented information of external origin determined by the organization to be necessary for the planning and operation of the quality management system shall be identified as appropriate, and be controlled. Documented information retained as evidence of conformity shall be protected from unintended alterations. Access can imply a decision regarding the permission to view the documented information only, or the permission and authority to view and change the documented information.

a. Guidelines: Annex A (A.6) ISO 9001:2015

As part of the alignment with other management system standards, a common clause on "documented information" has been adopted without significant change or addition. Where appropriate, text elsewhere in this international standard has been aligned with its requirements. Consequently, "documented information" is used for all document requirements.

Where ISO 9001:2008 used specific terminology such as "document" or "documented procedures", "quality manual" or a "quality plan", ISO 9001:2015 defines requirements to "maintain documented information".

Where ISO 9001:2008 used the term "records" to denote documents needed to provide evidence of conformity with requirements, this is now expressed as a requirement to "retain documented information". The organization is responsible for determining what documented information needs to be retained, the period of time for which it is to be retained and the media to be used for its retention.

A requirement to "maintain" documented information does not exclude the possibility that the organization might also need to "retain" that same documented information for a particular purpose, e.g. to retain previous versions of it.

Where ISO 9001:2015 refers to "information" rather than "documented information" (e.g. in clause 4.1: "The organization shall monitor and review the information about these external and internal issues"), there is no requirement that this information is to be documented. In such situations, the organization can decide whether or not it is necessary or appropriate to maintain documented information.

b. Guidelines: Clause 3.8.6 ISO 9000:2015

Information required to be controlled and maintained by an organization and the medium on which it is contained. Documented information can be in any format and media and from any source. Documented information can refer to: (a) The management system, including related processes; (b) information created in order for the organization to operate (documentation); (c) evidence of results achieved (records).

4.7.11 Key Requirements of Clause 7

Clause 7 requires that there are the right resources, people and infrastructure to meet the organizational goals. It requires an organization to determine and provide the necessary resources to establish, implement, maintain and continually improve the QMS. Simply expressed, this is a very powerful requirement covering all QMS resource needs and now covers both internal and external resources.

There are additional requirements to meet applicable statutory and regulatory requirements. The sub-clauses continue to cover requirements for infrastructure and environment for the operation of the processes. Monitoring and measuring includes resources such as personnel or training.

Organizational knowledge is a new requirement which deals with requirements for competence, awareness, and communication of the QMS. Personnel must not only be aware the quality policy, but they must also understand how they contribute to it and what the implications of non-conforming are.

There is key requirement to maintain the knowledge held by an organization to ensure conformity of products and services. This could include knowledge held by an individual for example, the intellectual property of an organization. Organizations are required to examine whether the current knowledge they have is sufficient when planning changes and whether any additional knowledge is required.

Finally there are requirements for "documented information". This is a new term, which replaces to "documents" and "records". ISO 900:2015 no longer mandates the need for documented procedures. However, it does specify the need to maintain and retain documented information, in order to give structure, clarity and evidence of the system being maintained and effective. Organizations need to determine the level of documented information necessary to control the QMS. This will differ between organizations due to size and complexity. Organizations still need to look at where documented information (e.g. processes, procedures, data, records) is critical for management systems and its effective operation. In line with the increased importance of information security in the organization, there is also greater emphasis on controlling access to documented information such as passwords.

4.8 OPERATION

4.8.1 Requirement: Clause 8.1 Operational Planning and Control

The organization shall plan, implement and control the processes needed to meet the requirements for the provision of products and services, and to implement the actions by: (a) Determining the requirements for the products and services; (b) establishing criteria for—(1) the processes, (2) the acceptance of products and services; (c) determining the resources needed to achieve conformity to the product and service requirements; (d) implementing control of the processes in accordance with the criteria; (e) determining, maintaining and retaining documented information to the extent necessary: (1) to have confidence that the processes have been carried out as planned; (2) to demonstrate the conformity of products and services to their requirements. The output of this planning shall be suitable for the organization's operations. The organization shall control planned changes and review the consequences of unintended changes, taking action to mitigate any adverse effects, as necessary. The organization shall ensure that outsourced processes are controlled.

Guidelines: This clause makes very clear statements about the importance of linking of operation to the critical elements where the critical processes and their interactions are determined. There are some additional requirements on control of changes, and also on control of outsourced processes. In construction the significance of quality is very important in operational planning and control, sincere efforts by all concerns must be given to impregnated quality right from this stage.

4.8.2 Requirement: Clause 8.2 Requirements for Products and Services

a. Customer Communication

Communication with customers shall include: (a) Providing information relating to products and services; (b) handling enquiries, contracts or orders, including changes; (c) obtaining customer feedback relating to products and services, including customer complaints; (d) handling or controlling customer property; (e) establishing specific requirements for contingency actions, when relevant.

b. Determining the Requirements for Products and Services

When determining the requirements for the products and services to be offered to customers, the organization shall ensure that: (a) The requirements for the products and services are defined, including (1) any applicable statutory and regulatory requirements; (2) those considered necessary by the organization; (b) the organization can meet the claims for the products and services it offers.

c. Review of the Requirements for Products and Services

The organization shall ensure that it has the ability to meet the requirements for products and services to be offered to customers. The organization shall conduct a review before committing to supply products and services to a customer, to include: (a) Requirements specified by the customer, including the requirements for delivery and post-delivery activities; (b) requirements not stated by the customer, but necessary for the specified or intended use, when known; (c) requirements specified by the organization; (d) statutory and regulatory requirements applicable to the products and services; (e) contract or order requirements differing from those previously expressed.

The organization shall ensure that contract or order requirements differing from those previously defined are resolved. The customer's requirements shall be confirmed by the organization before acceptance, when the customer does not provide a documented statement of their requirements. In some situations, such as internet sales, a formal review is impractical for each order. Instead, the review can cover relevant product information, such as catalogues.

The organization shall retain documented information, as applicable: (a) On the results of the review; (b) on any new requirements for the products and services.

d. Changes to Requirements for Products and Services

The organization shall ensure that relevant documented information is amended, and that relevant persons are made aware of the changed requirements, when the requirements for products and services are changed.

e. Guidelines

There must be a process to ensure the needs and expectations of customers are determined. This should include the determination of the intended product use and any statutory requirements that apply to the product in its intended market. Only once all requirements are identified can they be reviewed. Once determined, requirements need to be reviewed by the organization prior to its commitment to supply to ensure that they are understood, that any anomalies are resolved and that the organization has the ability to meet the requirements.

Communication needs to be planned to ensure that all necessary information is available when needed, from both external and internal sources. Documented information pertaining to communication is not specified, but typically it can include contracts, specifications, drawings, e-mails, letters, transmittals, meeting minutes, complaints, etc.

f. Guidelines: Clause 9.1.2 ISO 9001:2015

Customer feedback on delivered products and services can be an example of monitoring customer perception.

g. Guidelines: Clause 3.6.4 ISO 9000:2015

Requirement is defined as need or expectation that is stated, generally implied or obligatory. "Generally implied" means that it is custom or common practice for the organization and interested parties that the need or expectation under consideration is implied. A specified requirement is one that is stated.

4.8.3 Requirement: Clause 8.3 Design and Development of Products and Services

a. General

The organization shall establish, implement and maintain a design and development process that is appropriate to ensure the subsequent provision of products and services.

b. Design and Development Planning

In determining the stages and controls for design and development, the organization shall consider: (a) The nature, duration and complexity of the design and development activities; (b) the required process stages, including applicable design and development reviews; (c) the required design and development verification and validation activities; (d) the responsibilities and authorities involved in the design and development process; (e) the internal and external resource needs for the design and development of products and services; (f) the need to control interfaces between persons involved in the design and development process; (g) the need for involvement of customers and users in the design and development process; (h) the requirements for subsequent provision of products and services; (i) the level of control expected for the design and development process by customers and other relevant interested parties; (j) the documented information needed to demonstrate that design and development requirements have been met.

c. Design and Development Inputs

The organization shall determine the requirements essential for the specific types of products and services to be designed and developed. The organization shall consider: (a) functional and performance requirements; (b) information derived from previous similar design and development activities; (c) statutory and regulatory requirements; (d) standards or codes of practice that the organization has committed to implement; (e) potential consequences of failure due to the nature of the products and services. Inputs shall be adequate for design and development purposes, complete and unambiguous. Conflicting design and development inputs shall be resolved. The organization shall retain documented information on design and development inputs.

d. Design and Development Controls

The organization shall apply controls to the design and development process to ensure that: (a) the results to be achieved are defined; (b) reviews are conducted to evaluate the ability of the results of design and development to meet requirements; (c) verification activities are conducted to ensure that the design and development outputs meet the input requirements; (d) validation activities are conducted to ensure that the resulting products and services meet the requirements for the specified application or intended use; (e) any necessary actions are taken on problems determined during the reviews, or verification and validation activities; (f) documented information of these activities is retained. Design and development reviews, verification and validation have distinct purposes. They can be conducted separately or in any combination, as is suitable for the products and services of the organization.

e. Design and Development Outputs

The organization shall ensure that design and development outputs: (a) meet the input requirements; (b) are adequate for the subsequent processes for the provision of products and services; (c) include or reference monitoring and measuring requirements, as appropriate, and acceptance criteria; (d) specify the characteristics of the products and services that are essential for their intended purpose and their safe and proper provision. The organization shall retain documented information on design and development outputs.

f. Design and Development Changes

The organization shall identify, review and control changes made during, or subsequent to, the design and development of products and services, to the extent necessary to ensure that there is no adverse impact on conformity to requirements.

The organization shall retain documented information on: (a) Design and development changes; (b) the results of reviews; (c) the authorization of the changes; (d) the actions taken to prevent adverse impacts.

g. Guidelines

There must be a systematic approach to controlling design activities and product development. It will involve design planning which should include stages of design, review, verification and validation activities.

h. Guidelines: Clause 3.4.8 ISO 9000:2015

Design and development is defined as set of processes that transform requirements for an object into more detailed requirements for that object.

The requirements forming input to design and development are often the result of research and can be expressed in a broader, more general sense than the requirements forming the output of design and development. The requirements are generally defined in terms of characteristics. In a project there can be several design and development stages.

A qualifier can be applied to indicate the nature of what is being designed and developed (e.g. product design and development, service design and development or process design and development).

4.8.4 Requirement: Clause 8.4 Control of Externally Provided Processes, Products and Services

General

The organization shall ensure that externally provided processes, products and services conform to requirements. The organization shall determine the controls to be applied to externally provided processes, products and services when: (a) Products and services from external providers are intended for incorporation into the organization's own products and services; (b) products and services are provided directly to the customer(s) by external providers on behalf of the organization; (c) a process, or part of a process, is provided by an external provider as a result of a decision by the organization.

The organization shall determine and apply criteria for the evaluation, selection, monitoring of performance, and re-evaluation of external providers, based on their ability to provide processes or products and services in accordance with requirements. The organization shall retain documented information of these activities and any necessary actions arising from the evaluations.

a. **Type and Extent of Control**

The organization shall ensure that externally provided processes, products and services do not adversely affect the organization's ability to consistently deliver conforming products and services to its customers. The organization shall: (a) ensure that externally provided processes remain within the control of its quality management system; (b) define both the controls that it intends to apply to an external provider and those it intends to apply to the resulting output; (c) take into consideration—(1) The potential impact of the externally provided processes, products and services on the organization's ability to consistently meet customer and applicable statutory and regulatory requirements; (2) the effectiveness of the controls applied by the external provider; (d) determine the verification, or other activities, necessary to ensure that the externally provided processes, products and services meet requirements.

b. **Information for External Providers**

The organization shall ensure the adequacy of requirements prior to their communication to the external provider. The organization shall communicate to external providers its requirements for: (a) The processes, products and services to be provided; (b) the approval of—(1) products and services; (2) methods, processes and equipment; (3) the release of products and services; (c) competence, including any required qualification of persons; (d) the external providers' interactions with the organization; (e) control and monitoring of the external providers' performance to be applied by the organization; (f) verification or validation activities that the organization, or its customer, intends to perform at the external providers' premises.

c. Guidelines
The main aim of this requirement is to ensure that the purchased processes, products or services you require (e.g. components for your product) will ensure that you can meet your customer's requirements.

d. Guidelines: Annex A (A.8) ISO 9001:2015
All forms of externally provided processes, products and services have to suitably addressed through: (a) Purchasing from a supplier; (b) an arrangement with an associate company; (c) outsourcing processes to an external provider.

Outsourcing always has the essential characteristic of a service, since it will have at least one activity necessarily performed at the interface between the provider and the organization.

The controls required for external provision can vary widely depending on the nature of the processes, products and services. The organization can apply risk-based thinking to determine the type and extent of controls appropriate to particular external providers and externally provided processes, products and services.

e. Guidelines: Clause 9.6 ISO 9004:2018
Organizations procure externally supplied resources from a variety of providers. As these resources can impact both the organization and its interested parties, it is essential its relationships with external providers and partners are managed effectively. The organization and its external providers or partners are interdependent. The organization should seek to establish relationships that enhance the capabilities of itself and its providers or partners to create value in a manner that is mutually beneficial to all involved.

The organization should consider partnering if external providers have knowledge that the organization does not have, or to share the risks and opportunities associated with its projects (and the resulting profit or losses). Partners can be external providers of processes, products or services, technological and financial institutions, governmental and non-governmental organizations, or other interested parties.

In order to establish mutually beneficial relationships and to enhance the abilities of external providers and partners for managing activities, processes and systems, the organization should:

a. Share its mission and vision with them

b. Provide any necessary support

The managing of external providers should take into account the risks and opportunities associated with: (a) Internal facilities or capacity; (b) the technical capability to fulfill the requirements for products and services; (c) the availability of qualified resources

- The type and extent of controls needed for external providers
- Business continuity and supply chain aspects (e.g. high dependability on a single or limited number of providers)
- Environmental, sustainability and social responsibility aspects

4.8.5 Requirement: Clause 8.5 Production and Service Provision

a. Control of Production and Service Provision

The organization shall implement production and service provision under controlled conditions. Controlled conditions shall include, as applicable: (a) The availability of documented information that defines—(1) the characteristics of the products to be produced, the services to be provided, or the activities to be performed; (2) the results to be achieved; (b) the availability and use of suitable monitoring and measuring resources; (c) the implementation of monitoring and measurement activities at appropriate stages to verify that criteria for control of processes or outputs, and acceptance criteria for products and services have been met; (d) the use of suitable infrastructure and environment for the operation of processes; (e) the appointment of competent persons, including any required qualification; (f) the validation, and periodic revalidation, of the ability to achieve planned results of the processes for production and service provision, where the resulting output cannot be verified by subsequent monitoring or measurement; (g) the implementation of actions to prevent human error; (h) the implementation of release, delivery and post-delivery activities.

b. Identification and Traceability

The organization shall use suitable means to identify outputs when it is necessary to ensure the conformity of products and services. The organization shall identify the status of outputs with respect to monitoring and measurement requirements throughout production and service provision. The organization shall control the unique identification of the outputs when traceability is a requirement, and shall retain the documented information necessary to enable traceability.

c. Property Belonging to Customers or External Providers

The organization shall exercise care with property belonging to customers or external providers while it is under the organization's control or being used by the organization. The organization shall identify, verify, protect and safeguard customers' or external providers' property provided for use or incorporation into the products and services. When the property of a customer or external provider is lost, damaged or otherwise found to be unsuitable for use, the organization shall report this to the customer or external provider and retain documented information on what has occurred. A customer's or external provider's property can include materials, components, tools and equipment, premises, intellectual property and personal data.

d. Preservation

The organization shall preserve the outputs during production and service provision, to the extent necessary to ensure conformity to requirements. Preservation can include identification, handling, contamination control, packaging, storage, transmission or transportation, and protection.

e. Post-delivery Activities

The organization shall meet requirements for post-delivery activities associated with the products and services. In determining the extent of post-delivery activities that are required, the organization shall consider: (a) Statutory and regulatory requirements; (b) the potential undesired consequences associated with its products and services; (c) the nature, use and intended lifetime of its products and services; (d) customer requirements; (e) customer feedback.

Post-delivery activities can include actions under warranty provisions, contractual obligations such as maintenance services, and supplementary services such as recycling or final disposal.

f. Control of Changes

The organization shall review and control changes for production or service provision, to the extent necessary to ensure continuing conformity with requirements. The organization shall retain documented information describing the results of the review of changes, the person(s) authorizing the change, and any necessary actions arising from the review.

g. Guidelines

This requirement is aiming to ensure that your production activities and operations are planned and then conducted in a manner ensuring control. There are many different ways to achieve control and methods can include controlled processes, procedures, drawings, specifications, work instructions, quality plans, operating and process criteria. A business must have confidence in the ability of its process to consistently deliver and meet customer expectations.

h. Guidelines: Clause 3.6.13 ISO 9000:2015

Traceability is defined as ability to trace the history, application or location of an object. When considering a product or a service, traceability can relate to: (a) The origin of materials and parts; (b) the processing history; (c) the distribution and location of the product or service after delivery.

4.8.6 Requirement: Clause 8.6 Release of Products and Services

The organization shall implement planned arrangements, at appropriate stages, to verify that the product and service requirements have been met. The release of products and services to the customer shall not proceed until the planned arrangements have been satisfactorily completed, unless otherwise approved by a relevant authority and, as applicable, by the customer. The organization shall retain documented information on the release of products and services. The documented information shall include: (a) Evidence of conformity with the acceptance criteria; (b) traceability to the person(s) authorizing the release.

a. Guidelines: Clause 3.12.7 ISO 9000:2015

Release is defined as permission to proceed to the next stage of a process or the next process.

4.8.7 Requirement: Clause 8.7 Control of Nonconforming Outputs

The organization shall ensure that outputs that do not conform to their requirements are identified and controlled to prevent their unintended use or delivery. The organization shall take appropriate action based on the nature of the nonconformity and its effect on the conformity of products and services. This shall also apply to nonconforming products and services detected after delivery of products, during or after the provision of services. The organization shall deal with nonconforming outputs in one or more of the following ways: (a) Correction; (b) segregation, containment, return or suspension of provision of products and services; (c) informing the customer; (d) obtaining authorization for acceptance under concession. Conformity to the requirements shall be verified when nonconforming outputs are corrected. The organization shall retain documented information that—(a) describes the nonconformity; (b) describes the actions taken; (c) describes any concessions obtained; (d) identifies the authority deciding the action in respect of the nonconformity.

i. Guidelines

This requirement is intended to ensure that nonconforming product is prevented from further processing, use or delivery.

ii. Guidelines: Clause 3.7.5 ISO 9000:2015

Output is defined as result of process. Whether an output of the organization is a product or a service depends on the preponderance of the characteristics involved.

4.8.8 Key Requirements of Clause 8

This clause deals with the execution of the plans and processes that enable the organization to meet customer requirements and design products and services. The quality built in design stage is critical for the success of *in situ* construction activities as well as for project management. The clause places greater emphasis on the control of processes especially planned changes and review of the consequences of unintended changes, and mitigating any adverse effects as necessary. Most companies write work instructions and flowcharts to define and standardize their processes. It is important to follow ISO 9001 requirements for document control from beginning when writing these documents.

The ISO 9001:2015 standard acknowledges the trend toward greater use of subcontractors and outsourcing. This is demonstrated by the requirement to establish criteria for monitoring the performance of these parties in addition to keeping records used to establish selection criteria. The quality in supplier selection is of paramount importance for the success of construction project.

The clause now requires communication with regard to contingency actions where required and also the treatment of customer property. A new requirement for communicating with potential customers is also included, useful for bringing new offerings or solutions to the market.

There are more explicit requirements in terms of standards or codes of practice that the organization has committed to implement; internal and external resource needs for the design and development of products and services and finally the potential consequences of failure due to nature of products and services.

There is also new clause which covers post-delivery activities. This could include activities such as maintenance programmes or work carried out under warranty, and activities covering final disposal or recycling of the product. When determining the extent of these activities organizations must consider the risks associated with the product or service, customer requirements, customer feedback, and any statutory requirements.

4.9 PERFORMANCE EVALUATION

a. Requirement: Clause 9.1 Monitoring, Measurement, Analysis and Evaluation (General)

The organization shall determine: (a) What needs to be monitored and measured; (b) the methods for monitoring, measurement, analysis and evaluation needed to ensure valid results; (c) when the monitoring and measuring shall be performed; (d) when the results from monitoring and measurement shall be analyzed and evaluated. The organization shall evaluate the performance and the effectiveness of the quality management system. The organization shall retain appropriate documented information as evidence of the results.

b. Guidelines: Clause 10.4 ISO 9004:2018

The organization's performance should be evaluated from the viewpoint of the needs and expectations of interested parties. When deviations from the needs and expectations are found, the processes and their interactions that affects its performance should be identified and analyzed.

The organization's performance results should be evaluated against applicable objectives and their pre-determined criteria. Where objectives have not been achieved, the cause(s) should be investigated, with appropriate reviews of the deployment of organization's policies, strategy and objectives and the organization's managing of resources, as necessary. Similarly, when objectives have been exceeded, what made it possible should be analyzed in order to maintain the performance.

The results of evaluation should be understood by top management. Any identified performance failures should be prioritized for corrective action, based on the impact on the organization's policies, strategy and objectives.

Improvement achieved on the organization's performance should be evaluated from a long-term perspective. When the degree of improvement does not match the expected level, the organization should review the deployment of its policies, strategy and objectives for improvement and innovation, as well as the competencies and engagement of its people.

The organization's performance should be compared to established or agreed benchmarks. Benchmarking is a measurement and analysis methodology that an organization can use to search for best practices inside and outside the organization, with the aim of improving its performance and innovative practices. Benchmarking can be applied to policies, strategy and objectives, processes and their operations, products and services, or the organization's structure.

The organization should establish and maintain a methodology for benchmarking that defines rules for items, such as (a) the definition of the scope of the subject for benchmarking; (b) the processes for choosing benchmarking partner(s), as well as any necessary communications and confidentiality policies; (c) the determination of indicators for the characteristics to be compared and the data collection methodology to be used; (d) the collection and analysis of data; (e) the identification of performance gaps and the indication of potential improvement areas; (f) the establishment and monitoring of corresponding improvement plans; (g) the inclusion of gathered experience into the organization's knowledge base and learning process.

The organization should consider the different types of benchmarking practices, such as (a) internal benchmarking for activities and processes within the organization; (b) competitive benchmarking of performance or processes with competitors; (c) generic benchmarking by comparing strategies, operations or processes with unrelated organizations.

When establishing a benchmarking process, the organization should take into account that successful benchmarking depend on factors such as: (a) Support from top management (as it involves mutual knowledge interchange between the organization and its benchmarking partners; (b) the methodology used to apply benchmarking; (c) an estimation of benefits versus costs; (d) an understanding of the characteristics of the subject being investigated, in order to allow a correct comparison with the current situation in the organization; (e) implementing lessons learned to bridge any determined gaps.

4.9.1 Requirement: Clause 9.1.2 Customer Satisfaction

The organization shall monitor customers' perceptions of the degree to which their needs and expectations have been fulfilled. The organization shall determine the methods for obtaining, monitoring and reviewing this information. Examples of monitoring customer perceptions can include customer surveys, customer feedback on delivered products and services, meetings with customers, market-share analysis, compliments, warranty claims and dealer reports.

a. Guidelines: Clause 3.6.4 ISO 9000:2015

It can be necessary for achieving high customer satisfaction to fulfill an expectation of a customer even if it is neither stated nor generally implied or obligatory.

b. Guidelines: Clause 3.9.2 ISO 9000:2015

Customer Satisfaction can be defined as Customer's perception of the degree to which the customer's expectations have been fulfilled. It can be that the customer's expectation is not known to the organization, or even to the customer in question, until the product or service is delivered. It can be necessary for achieving high customer satisfaction to fulfill an expectation of a customer even if it is neither stated nor generally implied or obligatory.

Complaints are a common indicator of low customer satisfaction but their absence does not necessarily imply high customer satisfaction. Even when customer requirements have been agreed with the customer and fulfilled, this does not necessarily ensure high customer satisfaction.

c. Guidelines: Clause 4.1 ISO 9004:2018

The quality of an organization is the degree to which the inherent characteristics of the organization fulfill the needs and expectations of its customers and other interested parties, in order to achieve sustained success. It is up to the organization to determine what is relevant to achieve sustained success.

The organization should go beyond the quality of its products and services and the needs and expectations of its customers. To achieve sustained success, the organization should focus on anticipating and meeting the needs and expectations of its interested parties, with the intent of enhancing their satisfaction and overall experience.

The organization should apply all the quality management principles to achieve sustained success. Particular attention should be given to the principles of "Customer focus" and "Relationship management" to meet the needs and expectations of interested parties.

The needs and expectations of individual interested parties can be different, aligned to, or in conflict with those of other interested parties, or can change quickly. The means by which the needs and expectations of interested parties are expressed and met can take a wide variety of forms, for example, co-operation, negotiation, outsourcing, or by terminating an activity; consequently, the organization should give consideration to the interrelationships of its interested parties when addressing their needs and expectations. The composition of interested parties can differ significantly over time and between organizations, industries, cultures and nations.

4.9.2 Requirement: Clause 9.1.3 Analysis and Evaluation

The organization shall analyse and evaluate appropriate data and information arising from monitoring and measurement. The results of analysis shall be used to evaluate: (a) Conformity of products and services; (b) the degree of customer satisfaction; (c) the performance and effectiveness of the quality management system; (d) if planning has been implemented effectively; (e) the effectiveness of actions taken to address risks and opportunities; (f) the performance of external providers; (g) the need for improvements to the quality management system. Methods to analyse data can include statistical techniques.

a. Guidelines: Clause 10 ISO 9004:2018

The organization should establish a systematic approach to collect, analyze and review available information. Based on the results, the organization should use the information to update its understanding of its context, policies, strategy and objectives as needed, while also promoting improvement, learning and innovation activities. The available information should include data on: (a) Organization's performance; (b) the status of organization's internal activities and resources, which can be understood through internal audits or self-assessment; (c) changes in organization's external and internal issues and needs and expectations of the interested parties.

The organization should assess its progress in achieving its planned results against its mission, vision, policies, strategy and objectives, at all levels and in all relevant processes and functions. A measurement and analysis process should be used to monitor this progress, to gather and provide the information necessary for performance evaluation and effective decision making.

The selection of appropriate performance indicators and monitoring methods is critical for effective measurement and analysis of an organization. The steps for using performance indicators are: (a) Inventory of all processes; (b) select performance indicators and monitoring methods for processes; (c) measure, analyze and evaluate performance; (d) improve process.

The methods used for collecting information regarding performance indicators should be practicable and appropriate to the organization, such as (a) the monitoring and recording of process variables and product and service characteristic; (b) risk assessments of processes, products and services; (c) performance reviews, including on external providers and partners; (d) interviews, questionnaires and surveys on the satisfaction of interested parties.

Factors that are within the control of the organization and critical to its sustained success, should be subject to measurement and identified as key performance indicators (KPIs). These measurable KPIs should be: (a) Accurate and reliable, to enable the organization to set measurable objectives, monitor and predict trends, and take actions for improvement and innovation when necessary; (b) selected as a basis for making strategic and operational decisions; (c) suitably cascaded as performance indicators at relevant functions and levels within the organization, to support the achievement of top level objectives; (d) appropriate to nature and size of the organization, its products and services, processes and activities; (e) consistent with the strategy and objectives of the organization.

The organization should consider specific information relating to risks and opportunities when selecting KPIs. In addition, the organization should ensure that KPIs provide information to implement action plans when performance does not achieve the objectives, or to improve and innovate process efficiency and effectiveness. Such information should take into account elements such as (a) the needs and expectations of interested parties; (b) the importance of individual products and services to the organization; (c) the effectiveness and efficiency of processes; (d) the effectiveness and efficient use of resources; (e) financial performance; (f) compliance with applicable external requirements.

Analysis of the organization's performance should enable identification of issues, such as (a) insufficient and ineffective resources within the organization; (b) insufficient and ineffective competencies, organizational knowledge and inappropriate behaviour; (c) risks and opportunities that are not being sufficiently addressed by the organization's management systems; (d) weakness in leadership activities, including—(1) policy establishment and communication; (2) the managing of processes; (3) the managing of resources; (4) improvement, learning and innovation; (e) potential strengths that might need to be fostered with regard to leadership activities; (f) processes and activities showing outstanding performance that could be used as a model to improve other processes.

The organization should have a clear framework to demonstrate the interrelations between its leadership activities and their effects on the organization's performance. This can enable the organization to analyze the strengths and weaknesses of its leadership activities.

4.9.3 Requirement: Clause 9.2 Internal Audit

The organization shall conduct internal audits at planned intervals to provide information on whether the quality management system: (a) Conforms to—(1) the organization's own requirements for its quality management system; (2) the requirements of this international standard; (b) is effectively implemented and maintained.

The organization shall: (a) plan, establish, implement and maintain an audit programme (s) including the frequency, methods, responsibilities, planning requirements and reporting, which shall take into consideration the importance of the processes concerned, changes affecting the organization, and the results of previous audits; (b) define the audit criteria and scope for each audit; (c) select auditors and conduct audits to ensure objectivity and the impartiality of the audit process; (d) ensure that the results of the audits are reported to relevant management; (e) take appropriate correction and corrective actions without undue delay; (f) retain documented information as evidence of the implementation of the audit programme and the audit results.

a. Guidelines: Clause 3.13.1 ISO 9000:2015

Audit is defined as systematic, independent and documented process for obtaining objective evidence and evaluating it objectively to determine the extent to which audit criteria are fulfilled.

The fundamental elements of an audit include, the determination of conformity of an object according to a procedure carried out by personnel not being responsible for the object audited. An audit can be an internal audit (first party), or an external audit (second party or third party) and it can be a combined audit or a joint audit.

Internal audits, sometimes called first-party audits, are conducted by, or on behalf of, the organization itself for management review, and other internal purposes, and can form the basis for an organization's declaration of conformity. Independence can be demonstrated by the freedom from responsibility for the activity being audited.

External audits include those generally called second and third-party audits. Second party audits are conducted by parties having interest in the organization, such as customers, or by other persons on their behalf. Third-party audits are conducted by external, independent auditing organizations such as those providing certification/registration of conformity or government agencies.

b. Guidelines: Clause 10.5 ISO 9004:2018

Internal audits are an effective tool for determining the levels of conformity of the organization's management system to its selected criteria. They provide valuable information for understanding, analyzing and improving the organization's performance. Internal audit should assess the implementation, effectiveness and efficiency of the organization's management system. This can include auditing against more than one management system standard, as well as addressing specific requirements relating to interested parties, products, services, processes or specific issues.

To be effective, internal audits should be conducted in a consistent manner, by competent people, in accordance with the organization's audit planning. Audits should be conducted by the people who are not involved in the activity being examined, in order to give an independent view on what is being performed.

Internal auditing is an effective tool for identifying problems, nonconformities, risks and opportunities, as well as for monitoring progress on resolving previously identified problems and nonconformities. Internal auditing can also be focused on identification of good practices and on improvement opportunities.

The output of internal audits provide a useful source of information for:

- Addressing problems, nonconformities and risks
- Identifying opportunities
- Promoting good practices within the organization
- Increasing understanding of the interactions between processes.

Internal audit reporting usually contains information on conformity to the given criteria, nonconformities and improvement opportunities. Audit reporting is an essential input for management review. Top management should establish a process for reviewing all internal audit results, in order to identify trends that can acquire organization-wide corrective actions and opportunities for improvement.

The organization should use the results of other audits, such as second- and third-party audits, as feedback for corrective actions. It can also use them to monitor progress in the implementation of appropriate plans intended to facilitate the resolution of nonconformities, or for the implementation of identified opportunities for improvements.

4.9.4 Requirement: Clause 9.3 Management Review

Top management shall review the organization's quality management system, at planned intervals, to ensure its continuing suitability, adequacy, effectiveness and alignment with the strategic direction of the organization.

The management review shall be planned and carried out taking into consideration: (a) The status of actions from previous management reviews; (b) changes in external and internal issues that are relevant to the quality management system; (c) information on the performance and effectiveness of the quality management system, including trends in—(1) customer satisfaction and feedback from relevant interested parties, (2) the extent to which quality objectives have been met, (3) process performance and conformity of products and services, (4) nonconformities and corrective actions, (5) monitoring and measurement results, (6) audit results, (7) the performance of external providers; (d) the adequacy of resources; (e) the effectiveness of actions taken to address risks and opportunities (*refer* to clause 6.1); (f) opportunities for improvement.

The outputs of the management review shall include decisions and actions related to: (a) Opportunities for improvement; (b) any need for changes to the quality management system; (c) resource needs. The organization shall retain documented information as evidence of the results of management reviews.

a. Guidelines: Clause 3.11.2 ISO 9000:2015

Review is defined as determination of the suitability, adequacy or effectiveness of an object to achieve established objectives. Example: Management review, design and development review, review of customer requirements, review of corrective action and peer review. Review can also include determination of efficiency.

b. Guidelines: Clause 10.7 ISO 9004:2018

Reviews of performance measurement, benchmarking, analysis and evaluations, internal audits and self-assessments should be performed by appropriate levels and functions of the organization, as well as by top management. The reviews should be conducted at planned and periodic intervals, to enable trends to be determined and to evaluate the organization's progress towards achieving its policies, strategy and objectives. They should also address the assessment and evaluation of improvement, learning and innovation activities performed previously, including aspects of adaptability, flexibility and responsiveness in relation to the organization's mission, vision, values and culture.

The reviews should be used by the organization to understand the needs of adapting its policies, strategy and objectives. They should also be used to determine the opportunities for improvement, learning and innovation of the organization's managerial activities. The reviews should enable evidence-based decision making and establishment of actions to achieve desired results.

4.9.5 Key Requirements of Clause 9

Requirements for monitoring, measurement, analysis and evaluation are covered and you will need to consider what needs to be measured, methods employed, when data should be analyzed and reported on and at what intervals. Documented information that provides evidence of this must be retained.

There is now emphasis on directly seeking out information that relates to customer view the organization. Organizations must actively seek out information on customer perception. This can be achieved in number of ways including satisfaction surveys, analysis of market share, and through complaints logged.

There is now an explicit requirement that organization must show how the analysis and evaluation of this data is used, especially with regard to the need for improvements to the QMS.

Internal audits must also be conducted. There is additional requirements relating to defining the audit criteria and ensuring the results of the audits are reported to "relevant" management.

Management reviews are required with some additional requirements including the consideration of changes in external and internal issues that are relevant to the QMS. Documented information must be retained as evidence of management reviews.

4.10 IMPROVEMENT

4.10.1 Requirement: Clause 10.1 Improvement—General

The organization shall determine and select opportunities for improvement and implement any necessary actions to meet customer requirements and enhance customer satisfaction. These shall include: (a) Improving products and services to meet requirements as well as to address future needs and expectations; (b) correcting, preventing or reducing undesired effects; (c) improving the performance and effectiveness of the quality management system.

Examples of improvement can include correction, corrective action, continual improvement, breakthrough change, innovation and re-organization.

a. Guidelines: Clause 3.3.1 ISO 9000:2015

Improvement is defined as an activity to enhance performance.

b. Guidelines: Clause 11.2 ISO 9004:2018

Improvement is an activity to enhance performance. Performance can be related to a product or service, or to a process. Improving product or service performance or the management system can help the organization anticipate and meet the needs and expectations of interested parties and also increase economic efficiency. Improving processes can lead to increased effectiveness and efficiency, resulting in benefits such as cost, time and energy saving and reduce waste; in turn, this can lead to meeting the needs and expectations of interested parties more effectively.

Improvement activities can range from small-step continual improvements to significant improvements of the entire organization. The organization should define objectives for improving its products or services, processes, structure and management system, by using the results of analysis and evaluation of its performance.

Improvement processes should follow a structured approach. This methodology should be applied consistently for all processes. The organization should ensure that improvement becomes established as a part of organization's culture by: (a) Empowering people to participate in and contribute to successful achievement of improvement initiatives; (b) providing the necessary resources to achieve improvements; (c) establishing recognition systems for improvements; (d) establishing recognition systems for improving the effectiveness and efficiency of the improvement process; (e) engagement of top management in improvement activities.

c. Guidelines: Clause 11.4 ISO 9004:2018

Innovation should result in improvement leading to new or changed products or services, processes, market position, or performance, enabling realization or redistribution of value. Change in the organization's external and internal issues and the needs and expectations of interested parties could require innovation. To support and promote innovation the organization should: (a) Identify specific needs for innovation and encourage innovative thinking in general; (b) establish and maintain processes that allow for effective innovation; (c) provide the resources needed to realize innovative ideas.

Innovation can be applied at all levels of the organization, through changes in:

- Technology or products or services (i.e. innovations that not only respond to the changing needs and expectations of interested parties, but also anticipate potential changes in the organization and in the life cycles of products and services).
- Processes (i.e. innovation in the methods for production and service provision, or innovation to improve process stability and reduce variation).
- The organization (i.e. innovation in the constitution and the structure of the organization).
- The organization's management system (i.e. to ensure that competitive advantage is maintained and new opportunities are utilized, when there are emerging changes in the organization's context).
- The organization's business model (i.e. innovation in responding to distribution of value to customers or changing market position in accordance with interested parties needs and expectations).

4.10.2 Requirement: Clause 10.2 Nonconformity and Corrective action

When a nonconformity occurs, including any arising from complaints, the organization shall: (a) React to the nonconformity and, as applicable—(1) take action to control and correct it, (2) deal with the consequences; (b) evaluate the need for action to eliminate the cause(s) of the nonconformity, in order that it does not recur or occur elsewhere, by—(1) reviewing and analysing the nonconformity, (2) determining the causes of the nonconformity, (3) determining if similar nonconformities exist, or could potentially occur; (c) implement any action needed; (d) review the effectiveness of any corrective action taken; (e) update risks and opportunities determined during planning, if necessary; (f) make changes to the quality management system, if necessary. Corrective actions shall be appropriate to the effects of the nonconformities encountered.

The organization shall retain documented information as evidence of: (a) The nature of the nonconformities and any subsequent actions taken; (b) the results of any corrective action.

a. Guidelines: Clause 3.12.2 ISO 9000:2015

Corrective action is defined as action to eliminate the cause of a nonconformity and to prevent its recurrence. There can be more than one cause for a non-conformity. Corrective action is taken to prevent recurrence whereas preventive action is taken to prevent occurrence.

4.10.3 Requirement: Clause 10.3 Continual Improvement

The organization shall continually improve the suitability, adequacy and effectiveness of the quality management system. The organization shall consider the results of analysis and evaluation, and the outputs from management review, to determine if there are needs or opportunities that shall be addressed as part of continual improvement.

a. Guidelines: Clause 3.3.2 ISO 9000:2015

Continual improvement is defined as recurring activity to enhance performance. The process of establishing objectives and finding opportunities for improvement is a continual process through the use of audit findings and audit conclusions, analysis of data, management reviews or other means and generally leads to corrective action or preventive action.

4.10.4 Key Requirements of Clause 10

This clause starts with a new section that organizations should determine and identify opportunities for improvement such as improved processes to enhance customer satisfaction. There is also a need to actively look for opportunities to improve processes, products and services, and the QMS, especially with future customer requirements in mind.

Due to the new way of handling preventive actions, there are no preventive action requirements in this clause. However, there are some new corrective action requirements. The first is to react to the nonconformities, and take action, as applicable, to control and correct the nonconformities and deal with the consequences. The second is to determine whether similar nonconformities exist or potentially occur. The requirement for continual improvement has been extended to over the suitability and adequacy of the QMS as well as its effectiveness.

BIBLIOGRAPHY

1. A step by step guide on how to interpret each clause (nsi.org.uk)
2. clause_by_clause_explanation_of_iso_9001_2015_en.pdf (wordpress.com)
3. ISO 9000 Quality Management System—Fundamental and Vocabulary.
4. ISO 9001 Quality Management System—Requirements.
5. ISO 9001:2015 Requirements—Summary of Each Section (the9000store.com)
6. ISO 9004 Quality Management—Quality of an Organization—Guidance to achieve sustained success

5

Implementation Agencies for ISO 9001

OVERVIEW

Accreditation is the formal recognition by an authoritative body of the competence to work to specified standards. The apex role for ISO 9001 accreditation authorization is done by International Accreditation Forum (IAF) through Multilateral Recognitions Arrangements (MLA) for mutual recognition across the world. In India, Quality Council of India (QCI) is performing its pivotal role in the specific areas of accreditation as well as for quality promotion. As part of QCI, National Accreditation Board for Certification Bodies (NABCB) is signatory of IAF through MLA for ISO 9001. NABCB accredited certification bodies works in India. Other certification bodies also work in India which are accredited by other equivalent accreditation bodies working abroad. Bureau of Indian Standards (BIS) is national standard body accredited by NABCB for ISO 9001.

5.1 ACCREDITATION

5.1.1 Accreditation is a worldwide system for attesting competence and assuring equivalence of conformity assessment activities, which are testing, inspection and certification. It provides an independent, third party assessment to ensure that accredited conformity assessment bodies have the competence to perform specific tasks. The assessment covers verifying compliance of the management systems as well as requirements on impartiality and technical competence. The accreditation system facilitates trade and contributes to reducing technical barriers. That is, products and services can move across borders and be accepted without requirements of re-testing, re-certification or re-inspection. More and more governments and businesses are voluntarily choosing the accreditation route as they recognize the internal efficiencies that it brings and the opportunities to demonstrate product performance.

5.1.2 The key benefits of accreditation are: (a) Attestation of competence by authoritative body, (b) recognition of technical competence, (c) customer confidence and satisfaction, (d) minimizes risks, (e) avoids re-testing/inspection and reduces costs, (f) increased efficiency, (g) marketing advantage and increased business, (h) international recognition.

To understand accreditation it is equally important to understand what is meant by conformity assessment. Conformity assessment is the demonstration that what is being supplied or performed, actually meets the requirements specified or claimed. Conformity assessment can be applied to a product or a service, a process, a system, an organization or persons and includes activities such as testing, inspection, and certification. It is an objective based process having benefit of formal credibility recognition and reputational advantage. In a complex marketplace, where numerous goods and services are available, accreditation helps consumers to obtain greater confidence regarding performance.

5.2 ISO/IEC 17000 ACCREDITATION AND CERTIFICATION

a. Accreditation: The third party attestation related to conformity assessment body conveying formal demonstration of its competence to carry out specific conformity assessment tasks

b. Certification: Third party attestation related to products, processes, systems or persons.

It is important to understand the difference between accreditation body and accredited body:

a. Accreditation body (AB) is an authoritative body that performs accreditation. In some instances, its authority is derived from government.

b. Accredited body is an organization that provides certification, testing, calibration, inspection and other conformity assessment services. Accredited body related to certification body is generally called "Certification body (CB)". It is to be noted that all CBs are not accredited.

Certification represents a written assurance by a third party of the conformity of product, process or service to specified requirements. On the other hand, accreditation is the formal recognition by an authoritative body of the competence to work to specified standards.

India has a comprehensive system of product certifications governed by the Parliament of India at various times. These certifications are managed by various agencies, and hold various statutes before the law. Some of these marks are mandatory for such products to be manufactured or to be placed in Indian market while some of the marks hold only an advisory status. All the industrial standardization and industrial product certifications are governed by the Bureau of Indian standards, the national standards organizations of India, while standards for other areas are developed and managed by other government agencies.

The state enforced certification marks presently in India are: Agmark for all agricultural products, BIS hallmark for gold and silver jewellery, FSSAI for all food products, non polluting vehicle mark, geographical indications marks, etc.

5.3 INTERNATIONAL ACCREDITATION FORUM (IAF) AND INTERNATIONAL LABORATORY ACCREDITATION COOPERATION (ILAC)

5.3.1 Accreditation bodies (AB) are established in different countries as is beneficial to government and regulators, in complying regulatory and legislative responsibilities and acts as a flexible alternative to legislation. Accreditation facilitates trade and improves markets with more confidence in partners, goods and services. It is beneficial to business organizations for greater acceptance of their products and services; a tool for market access and the means to avoid costs of multiple testing, inspections or certifications. In addition, it minimises product failures or recalls and ensures more informed procurement selection. Accreditation bodies, which have been evaluated by peers as competent, sign arrangements that enhance the acceptance of products and services across national borders, thereby creating a framework to support international trade through the removal of technical barriers.

The International Accreditation Forum (IAF) is a global association of accreditation bodies, certification bodies, certification body associations and other organizations involved in conformity assessment activities in a variety of fields including management systems, products, services and personnel. Its website www.iaf.nu provides a range of information on certification bodies accreditation, as well as the location of its members including its Multilateral Recognition Arrangement (MLA) signatories worldwide. IAF promotes the worldwide acceptance, of certificates of conformity issued by certification bodies accredited by an accreditation body member.

The International Laboratory Accreditation Cooperation (ILAC) is the International authority on laboratory and inspection accreditation, with a membership consisting of accreditation bodies and affiliated organizations throughout the world. Its website www.ilac.org provides a range of information on laboratory and inspection accreditation, as well as the location of its members including its Mutual Recognition Arrangement (MRA) signatories worldwide. ILAC promotes the increased use and acceptance by industry as well as governments of the results from accredited laboratories and inspection bodies, including results from accredited organizations in the countries.

The primary role of IAF is to ensure that its accreditation body members only accredit bodies that are competent to do the work they undertake and are not subject to conflicts of interest. Accreditation bodies around the world, on being evaluated by their peers as competent, sign the IAF Multilateral Recognition Arrangement (MLA) or the ILAC Mutual Recognition Arrangement (MRA) that enhances the acceptance of products and services across national borders. Acceptance of accreditation body into the IAF MLA/ILAC MRA is dependent upon being successfully evaluated by peers from other accreditation bodies in accordance with relevant rules, standards and procedures. The purpose of these multilateral MLA/MRA is to facilitate mutual recognition of accredited certificates and reports between economies of signatory bodies and to create an international framework to support international trade through the removal of technical barriers.

A global initiative to raise the awareness of importance of accreditation, International Accreditation Forum (IAF) and the International Laboratory Accreditation Cooperation (ILAC), jointly marked 9th June as World Accreditation Day.

IAF has established mutual recognition arrangement, known as Multilateral Recognition Arrangement (MLA) reduces risk to business and its customers by ensuring that an accreditation certificate may be relied upon anywhere in the world. The IAF MLA consists of an agreement between accreditation bodies for mutual recognition of accreditation between signatories to the MLA.

Accreditations granted by IAF MLA signatories are recognised worldwide based on their equivalent accreditation programs. The IAF MLA mark can be used by accreditation bodies to demonstrate their status as a signatory to the IAF MLA. The IAF MLA provides government and regulatory agencies with a credible and robust framework on which to further develop and enhance Government to Government bilateral and multilateral international trade agreements. Accreditation and the IAF MLA help regulators meet their own responsibilities by providing a globally recognised system to accept accredited certification. The IAF MLA relies heavily on the MLAs of Recognised Regional Accreditation Groups such as the European Co-operation for Accreditation (EA), the Asia Pacific Accreditation Cooperation Incorporated (APAC) and the Inter American Accreditation Cooperation (IAAC), as it is these groups that perform the majority of the peer evaluation activity, not the IAF.

5.3.2 Categories of Membership to the International Accreditation Forum

a. Accreditation Body Members

IAF Signatories

Accreditation body members of IAF achieve IAF MLA signatory status after a full evaluation of their operations by a peer evaluation team, which is tasked to ensure that the applicant member complies fully with both the international standards and IAF documents. Once an accreditation body is a signatory of the IAF MLA, it is required to recognise and promote certificates issued by certification bodies accredited by all other signatories within the scope of the IAF MLA.

IAF Members that are not yet Signatories

Membership of IAF is open to accreditation bodies that conduct and administer programmes by which they accredit bodies for certification of quality systems, products, services, personnel, and environmental management systems, as well as other programmes of conformity assessment. Accreditation body members should declare their intention to join the IAF Multilateral Recognition Agreement (MLA) recognising the equivalence of other members' accreditations to their own.

b. Association Members

Association members are organisations or associations that represent a similar group of entities internationally or within an economy or region. These entities are associated with the programmes of IAF accreditation body members and they fully support IAF objectives.

c. Regional Accreditation Group Members

Regional accreditation group members consist of associations of accreditation bodies, and possibly other bodies, that cooperate within an identified geographic region to establish and maintain a multilateral recognition agreement based on a peer evaluation system, and represent the interests of accredited entities, industry, users and similar organisations that engage in, are subject to, make use of, accept or rely on conformity assessment results from bodies accredited by accreditation body members of IAF, and which support the purpose of IAF. Regional accreditation groups are invited to be represented in committees established to enhance cooperation between IAF and the regional accreditation groups.

5.3.3 International Accreditation Forum—Multilateral Recognition Arrangement (IAF-MLA) and its scope

a. General

Accreditation bodies, which have been evaluated by peers as competent, have signed an arrangement, the IAF Multilateral Recognition Arrangement (MLA), that enhances the acceptance of goods and services across national borders. The purpose of the MLA, is to ensure mutual recognition of accredited certification between signatories to the MLA, and subsequently acceptance of accredited certification in many markets based on one accreditation. Signatories must recognise and support acceptance of certificates issued by organisations accredited by all other signatories of the MLA, provided the certificates are issued within the scope of the IAF MLA signatory. This recognition and acceptance removes technical barriers to trade (TBT) by reducing redundant conformity assessment. Accreditations granted by IAF MLA signatories should be recognised worldwide based on their equivalent accreditation programs, reducing costs and adding value to business and consumers.

b. Scope

Currently there are four major scope of activities for the IAF MLA: Management systems certification, product certification, certification of persons, as also validation and verification. Certificates are 'equally reliable' because the conformity assessment bodies conform to the same standard.

c. Benefits

The IAF MLA provides governments and regulatory agencies with a credible and robust framework on which to further develop and enhance government to government bilateral and multilateral international trade agreements. It represents an internationally recognised 'stamp of approval' to demonstrate compliance against agreed standards and requirements. Consequently, risk is minimised, as decisions will be based on reliable certificates. Many organizations, such as government agencies, have recognised the importance of credible accreditation programs that are developed against internationally recognised standards. Accreditation and the IAF MLA help regulators meet their own legislated responsibilities by providing a globally recognised system to accept accredited certification. The long-term aim is the fully accepted use and recognition, by both public and private industries, of accredited certification, including certificates from other economies. In this way, the free-trade goal of "certified once, accepted everywhere" will be realised.

5.3.4 Benefits of Using an Accredited Certification Body

Selecting the right organisation to carry out the certification is very important. Choosing a certification body that has been accredited by an accreditation body that is a signatory to the IAF Multilateral Recognition Arrangement (MLA) has proved that it complies with best practice. It is competent to deliver a consistently reliable, and impartial and accurate service which meets the appropriate, internationally-recognised standard.

Using an accredited certification body can:

- De-risk the procurement by taking the guesswork out of choosing a certification body by giving confidence that one will get the service that closely meets the requirements.
- Win new business particularly since the use of accredited conformity assessment services is increasingly a stipulation of specifiers in both the public and private sector.
- Gain access to overseas markets since certificates issued by bodies that are accredited by an IAF MLA signatory are recognised and accepted throughout the world.
- Help to identify best practice since the certification body is required to have appropriate knowledge of the business sector.
- Offer market differentiation and leadership by showing to others credible evidence of good practice.
- Demonstrate due diligence in the event of legal action.
- Reduce paperwork and increase efficiency by reducing the need to re-audit the business.

5.3.5 Procedure for resolution of complaint about the operation of an IAF Member or a Certification Body

IAF treats any complaints with the utmost concern and will deal with them expeditiously and in confidence. In the first instance, complaints against a conformity assessment body should be lodged with that particular organisation. If the matter cannot be resolved within a satisfactory timeframe, the complainant has the right to refer the matter to the local accreditation body. If the complaint still cannot be resolved, then the complaint should be referred to the regional cooperation body (if applicable), and then to IAF.

Anyone wishing to submit a complaint should do so by emailing the relevant details to the IAF Corporate Secretary together with any necessary documentation required to substantiate the complaint.

IAF mechanism for dealing with complaints is available for information from IAF PR1:2004 Procedure for the investigation and resolution of complaints.

5.3.6 Membership of International Accreditation Forum

Accreditation body membership of IAF is open to bodies that conduct and administer programmes by which they do accreditation of agencies for certification of quality systems, products, services, personnel, environmental management systems as well as other programmes of conformity assessment. Application forms can be obtained by contacting iaf@iaf.nu.

5.3.7 Advantages of IAF Membership

IAF Membership offers several advantages for the organisation and its economy:

- The opportunity to become a signatory of the IAF MLA, thereby providing international recognition for the accredited bodies.
- The opportunity to learn from, and interact with, experienced accreditation bodies to assist with the development of the system.

- Interaction with other developing accreditation systems to share experiences and seek common solutions to problems.
- The opportunity also to represent, and inform, the constituents on important matters in the international conformity assessment arena.
- Participation in the IAF annual general assembly.
- Participation in IAF committees covering policy matters, technical issues, promotion and marketing, certification representation and development support for new accreditation systems.
- Access to the IAF intranet which provides access to the latest drafts of IAF documents (for comment and voting), a range of IAF resource material, and information from IAF committees.

5.3.8 ISO's Committee on Conformity Assessment (CASCO)

CASCO is the ISO committee that works on issues relating to conformity assessment. CASCO develops policy and publishes standards related to conformity assessment, although it does not perform conformity assessment activities itself. CASCO is made up of a number of groups that take care of different tasks. CASCO has produced a number of standards related to conformity assessment practices. Some of the examples are:

a. ISO 17020   Conformity assessment—requirements for the operation of various types of bodies performing inspection

b. ISO17021-1   Conformity assessment—requirements for bodies providing audit and certification of management systems—Part 1

c. ISO17021-2   Conformity assessment—requirements for bodies providing audit and certification of management systems—Part 2: Competence requirements for auditing and certification of environmental management systems

d. ISO 17021-3   Conformity assessment—requirements for bodies providing audit and certification of management systems—Part 3: Competence requirements for auditing and certification of quality management systems

e. ISO17021-10   Conformity assessment—requirements for bodies providing audit and certification of management systems—Part 10: Competence requirements for auditing and certification of occupational health and safety management systems

f. ISO17024   Conformity assessment—general requirements for bodies operating certification of persons

g. ISO17025   Conformity assessment—general requirements for the competence of testing and calibration laboratories

h. ISO17065   Conformity assessment—requirements for bodies certifying products, process and services.

5.4 QUALITY COUNCIL OF INDIA (QCI)

5.4.1 General

a. QCI was set up in 1997 jointly by the Government of India and Indian industry represented by the three premier industry associations, i.e. Associated

Chambers of Commerce and Industry of India (ASSOCHAM), Confederation of Indian Industry (CII) and Federation of Indian Chambers of Commerce and Industry of India (FICCI), to establish and operate national accreditation structure and promote quality through National Quality Campaign. QCI is also assigned the task of monitoring and administering the National Quality Campaign and to oversee effective functioning of the National Information and Enquiry Services.

b. QCI is registered as non-profit society with its own Memorandum of Association. QCI is governed by a Council of 38 members with equal representations of government, industry and consumers. The Chairman of QCI is appointed by the Prime Minister on recommendation of the industry to the government. The Department of Industrial Policy and Promotion, Ministry of Commerce and Industry, is the nodal ministry for QCI.

c. The QCI is having following boards for different types of accreditation programmes:

   i. NABCB: National Accreditation Board for Certification Bodies
   ii. NABET: National Accreditation Board for Education and Training
   iii. NABH: National Accreditation Board for Hospitals and Healthcare Providers
   iv. NABL: National Accreditation Board for Testing and Calibration Laboratories.

   In addition, it has an exclusive board for promotion of quality, i.e. National Accreditation Board for Quality Promotion (NBQP)

d. In the era of huge infrastructure development in the country, Quality Council of India (QCI) has taken an initiative to establish an independent third party voluntary "RMC Plant Certification Scheme" to assure quality in operations and processes of RMC plants in the country. This scheme was developed with the active participation and technical support from Ready Mixed Concrete Manufacturers' Association (RMCMA), a Mumbai based non-profit industry organization of leading Ready Mix Concrete (RMC) producers in India; Building Materials and Technology Promotion Council (BMTPC) under Ministry of Housing and Urban Poverty Alleviation, Government of India and various other stakeholders. The RMC Plants Certification Scheme has two options for certification, one being 'RMC Capability Certification' and other being 'RMC 9000+ Capability Certification', the latter ensuring compliance to the requirements of ISO 9001 also in addition to the QCI scheme requirements.

5.5 NATIONAL ACCREDITATION BOARD FOR CERTIFICATION BODIES (NABCB)

a. General

NABCB a constituent board of the Quality Council of India (QCI), is the national accreditation body, which provides accreditation to certification as well as inspection bodies in accordance with ISO Standards, international requirements/ guidelines and NABCB accreditation criteria. NABCB is internationally recognized and represents the interests of Indian industry at the international forums under the aegis of the International Accreditation Forum (IAF) and the International Laboratory Accreditation Cooperation (ILAC) and its recognized regional bodies in the Asia Pacific region. The Pacific Accreditation Cooperation (PAC) has sole objective of promoting acceptance of India's certifications internationally by becoming a signatory to the Multilateral Recognition Arrangement of these bodies after undergoing successful international peer evaluations. NABCB is a member of IAF and PAC as well as signatory to their MLAs for Quality Management Systems (QMS), Environmental Management Systems (EMS), Product Certification, Food Safety Management System (FSMS), Information Security Management Systems (ISMS). Thus NABCB's accreditations are internationally acceptable. NABCB maintains a list of certification bodies accredited by it and the same is available at its website www.nabcb.qci.org.in.

b. Objectives of NABCB

The objective of NABCB is to establish and offer accreditation schemes, based on internationally accepted standards, for certification bodies and inspection bodies engaged in providing services of system certification (ISO 9001, ISO 14001, etc.) product certification and inspection. The criterion of accreditation adopted by NABCB is in line with the international requirements and to foster improvement in the quality of certification process with the support of certification bodies. NABCB remains impartial in decision making on criteria and process of accreditation and conducts its business professionally, independent of undue influence of any stakeholders.

c. Need of Accreditation

Accreditation is the recognized mechanism accepted by World Trade Organization/ Technical barriers to trade (WTO/TBT) agreements for establishing equivalence of certification/inspection schemes operated in different countries as also the test results of various laboratories.

d. Structure of NABCB

Chairman of the Board is nominated by Chairman of Quality Council of India. The other members are nominees of Ministry of Commerce, Department of Industrial Policy and Promotions, CII, FICCI and ASSOCHAM (industry associations), Bureau of Indian standards, nominees of association of certification bodies, nominee from consulting organizations and nominee of consumer bodies (11 members including Chairman). There can be other invitee members.

e. Accreditation Schemes Offered by NABCB

NABCB offers accreditation in the areas of Management System Certification [based on ISO 9001 (QMS), ISO 14001 (EMS)], product certification and inspection bodies.

f. NABCB's Finances and Fee Structure for Accreditation

NABCB gets its finances through the services offered and from no other sources. The fee structure is decided by NABCB Board from time to time. The current fee structure is a part of the application pack and is also available on request.

g. Worldwide Recognition of NABCB Accreditation

NABCB is a member of the Asia Pacific Accreditation Cooperation (Association of Accreditation Bodies in the Asia-Pacific Region) and International Accreditation Forum (Association of Accreditation Bodies worldwide). NABCB is also a signatory to the IAF MLA (Multilateral Arrangement for Mutual Recognition). This status has been achieved after undergoing an assessment by peers from other countries. Thus NABCB is empowered to resolve complaint about certification

Body. Each Certification body has a procedure to resolve the complaints, disputes and appeals. The organization should first write to the certification body for resolution of the complaint, dispute or appeal. In case the certification body is not able to satisfy the company raising complaint, dispute or appeal, then the organization can approach the respective accreditation body with complete details of the matter for redressal. If NABCB is approached with the complaint, it will be forwarded to the respective accreditation body if the certification body referred in the complaint is not accredited by NABCB.

h. Complaint about ISO Certified Company and role of NABCB

Each of the certified company has procedure to address the complaints of the customers for redressal of complaints related to them.

5.6 BUREAU OF INDIAN STANDARDS (BIS)

BIS is the National Standard Body of India established for the harmonious development of the activities of standardization, marking and quality certification of goods. It came into existence, through an act of parliament dated 26 November 1986 taking over the functions of ISI (Indian Standards Institution). Through this change over, the government envisaged building a climate for quality culture and consciousness and greater participation of consumers in formulation and implementation of national standards.

BIS also operates product and management systems certification. BIS has been operating Management Systems Certification Scheme since 1991. Initially, BIS started the scheme with Quality Management System Certification (IS/ISO 9001) and over the years it has gradually expanded its activities to various other management systems like Environmental Management System IS/ISO 14001, Occupational Health and Safety Management System IS/ISO 45001. BIS is accredited by National Accreditation Board for Certification Body (NABCB) for various management system certification schemes.

5.7 PRIVATE CERTIFICATION BODIES AND AWARD OF ISO CERTIFICATE

5.7.1 NABCB under Quality Council of India is providing accreditation for ISO certification bodies. The accreditation is part of an international network managed by the IAF. Accreditation bodies of other countries also works in India under the network of IAF. These bodies are also allowed to accredit certification bodies in India, besides NABCB. There are large numbers of certification bodies in India, to carry-out certification at competitive price.

The organizations need to make a wise choice to get maximum benefit out of certification, both internally (having better systems) and externally (to ensure customer confidence as well as international equivalence).

5.7.2 Audits and certifications to Management system standards (MSS) are a major asset to international trade and essential to the long-term health of the world economy. Attesting the conformity to management system standards, certificates are widely used in global markets to establish confidence between business partners and between organizations and their customers, to qualify suppliers in supply chains, and as a requirement to tender for procurement contracts. Certification bodies that use current ISO/IEC 17021–1:2015 Conformity assessment—requirements for bodies providing audit and certification of management systems, Part 1: Requirements are able to ensure competent audit teams, with adequate resources, following a consistent process and reporting audit results in a consistent manner. It relates to the competence of certification bodies themselves and their auditors. It applies to the auditing and certification of all types of management systems, in order to increase their value to public- and private-sector organizations worldwide. It also helps to create confidence among regulators, consumers, suppliers and other stakeholders that certificates granted by one certification body is effectively equivalent to that offered by another. ISO/IEC 17021–1:2015 focus more on how certification services are delivered by a certification body, and as such, the improvements are intended to:

a. Bolster the effectiveness of operational and organization control by certification bodies of remote offices, regardless of their organizational structure.

b. Enhance an additional risk management approach.

c. Define audit time and audit duration, and then focus requirements for justification on audit duration.

5.7.3 ISO/IEC 17021–1:2015 contains principles and requirements for the competence, consistency and impartiality of bodies providing audit and certification of all types of management systems. Certification of management systems is a third-party conformity assessment activity and bodies performing this activity are therefore third-party conformity assessment bodies. The accreditation of certification bodies is carried out based on following principles:

a. **Impartiality:** A certification body is required to provide confidence that the certification is based on principle of impartiality.

b. **Competence:** A certification body is required to provide confidence that the certification process is carried out by competent personnel.

c. **Responsibilities:** The certification body has the responsibility to grant certification if there is sufficient objective evidence of conformity of ISO standard requirements and not to grant certification if evidences are insufficient. Likewise, the certified client has the responsibility for consistently achieving the intended results of implementation of management system standard and conformity with the requirements for certification.

d. **Openness:** A certification body needs to gain confidence in the integrity and credibility of certification by disclosure of appropriate information about certification process and about status of certification of any organization.

e. **Confidentiality:** A certification body need not disclose any confidential information obtained to adequately assess conformity to requirements for certification.

f. **Responsiveness to complaints:** Parties that rely on certification should have confidence that their valid complaints will be appropriately addressed by the certification body. Confidence in certification activities is safeguarded when complaints are processed appropriately.

g. **Risk-based approach:** Certification bodies need to take into account the risks associated certification such as risks associated with objectives of audit, sampling used in audit process, real and perceived impartiality, legal, regulatory and liability issues, etc.

5.7.4 Certification bodies (CBs) can develop their own documentation for quality. Quality manual including policy manual, procedure manual and auditor's handbook. The policy manual can have the objective of the CB and can include: (a) Distribution list of procedure manual, (b) amendment record sheet, (c) all services provided by the CB, (d) define an appropriate administrative structure for activities, function to clear as per requirements of accreditation body, (e) organization structure, (f) certification personnel requirements, (g) certification and surveillance audit procedure, (h) internal and periodic review and (i) auditor's agreement and undertaking of confidentiality.

Likewise, the certification body has to develop procedure manual which can include: (a) Procedure for handling enquiries and complaints, (b) planning and conduct of assessments, (c) issue, change and cancellation of certificates, (d) internal audit and periodic reviews and (e) selection and assignment of auditors.

Auditor's Handbook of Certification Body ought to have: (a) Certification procedures, (b) code of conduct, (c) auditor responsibility, (d) pre-audit visit, (e) pre-assessment and initial assessment visits and (f) reporting.

BIBLIOGRAPHY

1. Accreditation-https://www.unido.org/sites/default/files/2016-10/SDG_ACCREDITATION_ BROCHURE__2__0.pdf
2. Bureau of Indian Standards-https://bis.gov.in/
3. https://www.iso.org/news/2015/06/Ref1972.html
4. International Accreditation Forum-https://www.iaf.nu/
5. International Organization for Standardization-https://www.iso.org/resources-for-conformity-assessment.html and https://www.iso.org/news/ref2250.html.
6. ISO CASCO Standards-https://www.iso.org/casco.html
7. National Accreditation Board for Certification Bodies-http://nabcb.qci.org.in/
8. Quality Council of India-http://qcin.org/

Quality Management System Auditing

OVERVIEW

ISO 9000 standards emphasize on the importance of QMS audit and review of its results by management. However, there is no requirement of documented procedure for Internal Audit in ISO 9001:2015. The terms and definitions related to QMS audit are given in ISO 9000:2015. The concepts of the class "Audit" and related concepts are explained in the form of concept diagram in ISO 9000:2015 standard. The terms and definitions related to audit and their relationships can be clearly understood through it.

ISO 19011 standard provides guidelines for auditing management systems which cover guidance on management of audit programme, on the planning and conducting of management system audits, as well as on the competence and evaluation of an auditor and an audit team. ISO 19011 standard is useful as guidance to internal audits of ISO 9001:2015 QMS. Auditing is characterized by reliance on a number of principles. These principles should help to make the audit an effective and reliable tool in support of management policies and controls, by providing information on which an organization can act in order to improve its performance.

The seven principles and processes of auditing covered in ISO 19011:2018 are the same for auditing of different management systems developed on ISO Annex SL framework basis, therefore, now two or more management systems of different disciplines can be audited together.

In present scenario, technology has also influenced auditing too. Auditing of a virtual location is sometimes referred to as virtual auditing; such audits are conducted when an organization performs work or provides a service using an online environment allowing persons irrespective of physical locations to execute processes (e.g. company intranet, a "company cloud"). A virtual audit follows the standard audit process while using technology to verify objective evidence; the technology is used to gather information, interview an auditee, etc. when "face-to-face" methods are not possible or desired.

The use of "process approach" is a requirement for all ISO management system standards in accordance with ISO/IEC Directives, Part 1, Annex SL. Auditors should understand that auditing a management system is auditing an organization's processes and their interactions in relation to one or more management system standard(s).

The audit is completed when all planned audit activities have been carried out, or as otherwise agreed with the audit client.

6.1 NECESSITY OF AUDIT

Audit is used as a tool to monitor and determine the health of the management system implemented in the organization. Management System Audit is a positive activity to gather information on whether the management system conforms to the requirements of chosen ISO standard and to the requirements of the organization.

The information from audit results can provide input to the analysis aspect of business planning and helps top management to evaluate the need for improvement or corrective action. An audit should not be confused with surveillance or inspection activities performed for the sole purpose of process control or product acceptance.

By using process audit, management can measure the suitability, adequacy and effectiveness and continuous improvement of its own QMS as well as the capability of existing and prospective external providers that support the organization. The QMS audit provides objective evidence concerning the need for reduction, elimination, and prevention of nonconformity within the organization and its external provider as well.

6.2 PRINCIPLES OF AUDITING: ISO 19011:2018 STANDARD

Auditing relies upon a set of principles to help make an audit an effective and reliable tool. Adherence to these principles is a prerequisite for providing audit conclusions that are relevant and sufficient, and for enabling auditors, working independently from one another; to reach similar conclusions in similar circumstances. The principles are:

  a. **Integrity:** The foundation of professionalism
    – Auditors and the individual(s) managing an audit programme should perform their work ethically, with honesty and responsibility.
    – Only undertake audit activities if competent to do so;
    – Perform their work in an impartial manner, i.e. fair and unbiased in all their dealings
    – Be sensitive to any influences that may be exerted on their judgment while carrying out an audit.

  b. **Fair presentation:** The obligation to report truthfully and accurately.

    Audit findings, audit conclusions and audit reports should reflect truthfully and accurately the audit activities. Significant obstacles encountered during the audit and unresolved diverging opinions between the audit team and auditee should be reported. The communication should be truthful, accurate, objective, timely, clear and complete.

  c. **Due professional care:** The application of diligence and judgment in auditing.

    Auditors should exercise due care in accordance with the importance of task they perform and the confidence placed in them by the audit client and other interested parties. An important factor in carrying out their work with duo professional care is having the ability to make reasoned judgments in all audit situations.

  d. **Confidentiality:** Security of information

    Auditors should exercise discretion in the use and protection of information acquired in the course of their duties. Audit information should not be used inappropriately for personal gain by the auditor or audit client, in a manner detrimental to the legitimate interests of the auditee. This concept includes the proper handling of sensitive or confidential information.

e. **Independence:** The basis for the impartiality of the audit and objectivity of the audit conclusions.

Auditors should be independent of the activity being audited wherever practicable, and should in all cases act in a manner that is free from bias and conflict of interest. For internal audits, auditors should be independent from the function being audited if practicable. Auditors should maintain objectivity throughout the audit process to ensure that the audit findings and conclusions are based only on the audit evidence.

For small organization, it may not be possible for internal auditors to be fully independent of the activity being audited, but every effort should be made to remove bias and encourage objectivity.

f. **Evidence-based approach:** This is a rational method for reaching reliable and reproducible audit conclusions in a systematic audit process

Audit evidence should be verifiable. It should in general be based on samples of the information available, since an audit is conducted during a finite period of time and with finite resources. An appropriate use of sampling should be applied, since this is closely related to the confidence that can be placed in the audit conclusions.

g. **Risk-based approach:** The risk-based approach should substantially influence the planning, conducting and reporting of audits in order to ensure that audits are focused on matters that are significant for the audit client, and for achieving the audit programme objectives.

6.3 AUDIT DEFINITION AND ITS TYPES

Audit is defined as "systematic, independent and documented process for obtaining objective evidence and evaluating it objectively to determine the extent to which the audit criteria are fulfilled". An audit can be an internal audit (first party), or an external audit (second party or third party).

6.3.1 Internal Audit

Internal audit, sometimes called first-party audit, is a mandatory requirement of ISO 9001 standard and most important of all audits. Internal audit is self-audit by the organization and generally conducted by its own auditors and can form the basis for an organization's self-declaration of conformity.

Internal audit requires the organization to look at its own system, processes, procedures and activities in order to determine whether they are adequate and compliant. It provides top management with the information on whether quality policy and objectives are met and if the system is suitable and effective. The information attained from the internal audit becomes a source for periodic management review. The internal audit programme shall be established which takes into consideration the importance of processes concerned, and the results of previous audits. It is likely that key or high risk areas are audited more frequently.

6.3.2 External Audit

External audits include those generally called second- and third-party audits. Second party audits are conducted by parties having an interest in the organization, such as customers, or by other persons on their behalf. The audit conducted by organization on their external providers or other external interested parties is also called second-party audit. This type of audit is of paramount importance in construction and is generally undertaken in an unorganized manner, not strictly as per the ISO Management System Standards.

Third-party audits are conducted by external, independent auditing organizations such as those providing certification/registration of conformity or governmental agencies.

Often it happens that external auditors are able to find gaps and improvement areas where internal audit portrays a highly satisfactory picture of QMS.

6.4 INTERNAL AUDITORS

Internal auditors are selected, trained and coached well to conduct a useful audit. The senior management can create an environment where internal audits are valued and taken seriously by all. Compliance verification is the basic requirement but as time progresses auditors need to go beyond that to keep QMS really adding new improvements. The internal auditors must be competent enough to bring together related concepts within a standard even if they appear on different clauses of standard. Audits are required to be conducted by trained and qualified members of the organization or by hired professional auditors. There are advantages to both options. The important consideration is which gives the maximum information. One question often asked is whether a special knowledge of the area activity to be audited is required. In theory the response is no, because the auditor should be looking for objective evidence based on the requirements of the standard and provided by the quality management system and conformity to it. In practice, general knowledge is preferred to assist in the analysis of data acquired and in the formation of a judgement. The auditor must be independent of the area being audited and not have direct responsibility for it.

Most organizations that are certified to ISO 9001 standard use organization's internal auditors who do internal audits on need based requirement. Training and conducting audits is such a small amount of their job that they never have time to tune and improve their audit skills. Employees often get promotions, get busier, or may even leave the organization. All of these circumstances cause organizations to be in a constant internal auditor training mode, leading to ineffective audits costing more each year, which normally are not resulting in a positive result.

When experienced, trained staffs conducts audits for a living, which is well trained, and very knowledgeable about improvement methods and techniques, then the internal audit will result in positive outcome. The main reason internal audits are not effective because they are being carried out by auditors as a check-list that is designed against quality management standard and not attuned to business appropriately.

Auditors should apply professional judgement during the audit process and avoid concentrating on the specific requirements of each clause of the standard at the expense of achieving the intended outcome of the management system.

Auditors should focus on intended result of the management system throughout the audit process. While processes and what they achieve are important, the result of the management system and its performance is what counts.

6.5 COMPETENCE

6.5.1 Competence of Individual(s) Managing Audit Programme

The individual(s) Managing Audit Programme should have necessary competence to manage the programme and its associated risks and opportunities and external and internal issues effectively and efficiently, including knowledge of:

a. Audit principles, methods and processes.
b. Management system standards, other relevant standards and references/guidance documents.
c. Information regarding auditee and its context (e.g. external/internal issues, relevant interested parties and their needs and expectations, business activities, products, services, and processes of auditee).
d. Applicable statutory and regulatory requirements and other requirements relevant to the business activities of the auditee.

As appropriate, knowledge of risk management, project and process management, and information and communication technology should be considered.

The individual(s) managing the audit programme should engage in appropriate continual development activities to maintain the necessary competence to manage the audit programme and establish suitable mechanism for the continual evaluation of the performance of the auditors.

6.5.2 Competence of Auditors

Confidence in audit process and the ability to achieve its objectives depends on the competence of those individuals who are involved in performing audits. Competence should be evaluated regularly through a process that consider personal behaviour and the ability to apply the knowledge and skills gained through education, work experience, auditor training and audit experience. This process should take into consideration the need of audit programme and its objectives.

Auditors should develop, maintain and improve their competence through continual professional development and regular participation in audit. This can be achieved through means such as additional work experience, training, private study, coaching, attendance at meetings, seminars and conferences or other relevant activities. Auditor should possess the knowledge and skills necessary to achieve the intended results of the audits they are expected to perform. The auditor's competence can be generic and also a level of discipline and sector-specific knowledge and skills. Auditors should exhibit professional behaviour during the performance of audit activities. Desired professional behaviour should include being:

a. Ethical, i.e. fair, truthful, sincere, honest and discreet.
b. Open-minded, i.e. willing to consider alternative idea or point of view.
c. Diplomatic, i.e. tactful in dealing with individuals.
d. Observant, i.e. actively observing physical surroundings and activities.

e. Perceptive, i.e. aware of and able to understand the situations.

f. Versatile, i.e. able to readily adapt to different situations.

g. Tenacious, i.e. persistent and focused on achieving objectives.

h. Decisive, i.e. able to reach timely conclusions based on logical reasoning and analysis.

i. Self-reliant, i.e. able to act and function independently while interacting effectively with others.

j. Able to act with fortitude, i.e. able to act responsibly and ethically, even though these actions may not always be popular and may sometimes result in disagreement or confrontation.

k. Open to improvement, i.e. willing to learn from situations.

l. Culturally sensitive, i.e. observant and respectful to the culture of the auditee.

m. Collaborative, i.e. effectively interacting with others, including audit team members and the auditee's personnel.

Auditor competence can be acquired using a combination of the following:

a. Successfully completing training programmes that cover generic auditor knowledge and skills. The knowledge and skill of auditor is of vital importance for success of ISO MSS audits.

b. Experience in relevant technical, managerial or professional position involving the exercise of judgement, decision making, problem solving and communication with managers, professionals, peers, customers and other relevant interested parties.

c. Education/training and experience in specific management system discipline and sector that contribute to the development of overall competence.

d. Audit experience acquired under supervision of an auditor competent in the same discipline.

6.6 MANAGING AN AUDIT PROGRAMME

ISO 9001:2015 requires organization's to plan, establish, implement and maintain an audit programme(s) including the frequency, methods, responsibilities, planning requirements and reporting, which shall take into consideration the importance of processes concerned, changes affecting the organization, and the results of previous audits.

The common objective of internal audit programme is to determine the continuing suitability, adequacy and effectiveness of the management system. The other examples of audit programme objectives should be to:

- Identify opportunities for the improvement of a management system and its performance.
- Conform to all relevant requirements, e.g. statutory and regulatory requirements
- Obtain and maintain capability of an external provider
- Evaluate the compatibility and alignment of the management system objectives with the strategic direction of the organization.

The planning of internal audit programmes and, in some cases programmes for auditing external providers, can be arranged to contribute to other objectives of the organization. The individual(s) managing the audit programme should ensure that the integrity of the audit is maintained and there is no undue influence exerted over the audit personnel.

Competent individuals should be assigned to manage the audit programme. Auditors-in-training may be included in the audit team, but should participate under the direction and guidance of an accomplished auditor. If the necessary competence is not displayed by the auditors in the audit team, technical experts with additional competence should be made available to support the team.

Audit programme can include audits addressing one or more management system standards or other requirements, conducted either separately or in combination (combined audit). In case of multiple locations/sites (e.g. different countries), or where important functions are outsourced and managed under the leadership of another organization, particular attention should be paid to the design, planning and validation of the audit programme.

The implementation of audit programme should be monitored and measured on an on-going basis to ensure its objectives have been achieved. The audit programme should be reviewed in order to identify needs for changes and possible opportunities for improvements.

There are risks and opportunities related to the context that can be associated with the audit programme and can affect the achievement of its objectives. Such risks are required to be addressed appropriately. Some of the examples are:

- Allowing insufficient time for audit
- Insufficient overall competence to conduct audit effectively.
- Ineffective internal communication of audit programme.
- Ineffective coordination of audits within the audit programme
- Ineffective determination of the necessary documented information
- Failure to adequately protect audit records to demonstrate audit programme effectiveness.
- Ineffective monitoring of audit programme outcome.

The individual audit objectives, scope and criteria to be consistent with overall audit programme objectives. The audit scope should be consistent with audit programme and audit objectives. It includes such factors as locations, functions, activities and processes to be audited, as well as the time period covered by the audit.

The audit criteria are used as a reference against which conformity is determined. These may include one or more of the following: Applicable policies, processes, procedures, performance criteria including objectives, statutory and regulatory requirements, management system requirements, information regarding the context and the risks and opportunities as determined by the auditee (including relevant external/internal interested parties requirements), sector codes of conduct or other planned arrangements.

In the event of any change to the audit objectives, scope or criteria, the audit programme should be modified if necessary and communicated to interested parties as appropriate.

The individual(s) managing the audit programme should ensure that audit records are generated, managed and maintained to demonstrate the implementation of the audit programme; should review the audit programme to assess whether its objectives have been achieved; establish suitable mechanism for the continual evaluation of the performance of auditors.

6.7 CONDUCTING AUDIT

The task involves: Initiating audit, preparing audit activities, conducting audit activities, preparing and distributing audit report, completing audit and conducting audit follow-up.

The auditor should request access to relevant information for planning purposes including information on the risks and opportunities the organization has identified and how they are addressed. Also, the auditor should also determine applicable statutory and regulatory requirements and other requirements relevant to the activities, processes, products and services of the auditee. To achieve the audit objectives, sufficient and appropriate information for planning and conducting audit, adequate cooperation from auditee, adequate time and resources is required.

Preparing audit activities require performing review of documented information, audit planning by adopting risk-based approach to planning, assigning work to audit team and preparing documented information for the audit.

The auditor should understand auditee's operation and applicable work documents before conducting of audit. The audit planning should also consider any follow-up actions from previous audit or other sources, e.g. lessons learned, project reviews. The scale and content of the audit planning can differ, for example, between initial and subsequent audits. Audit planning should be sufficiently flexible to permit changes which can become necessary as the audit activities progresses. The auditor should review auditee's relevant documented information to determine the conformity of the system.

Conducting audit includes activities like conducting opening and closing meeting with the auditee, communication during audit, audit information availability and access, reviewing documented information during audit, collecting and verifying information, generating audit findings, determining audit conclusions, preparing and distributing audit report, completing audit and conducting audit follow-up.

Audit findings can indicate conformity or non-conformity with audit criteria. Nonconformities and their supporting audit evidence should be recorded. They should be reviewed with the auditee in order to obtain acknowledgement that audit evidence is accurate and that the nonconformity is understood.

The outcome of the audit can, depending on the audit objectives, indicate the need for corrections or for corrective actions, or opportunities for improvement. Such actions are usually decided and undertaken by the auditee within an agreed timeframe. The completion and effectiveness of these actions should be verified. This verification may be part of a subsequent audit.

6.8 ISO 9001:2015 AUDIT

ISO 9001:2015 has number of new requirements which will change focus of audit. One of the major changes is that it brings quality management and continual improvement into the heart of an organization. This means that the QMS must be aligned with the strategic direction of the organization. There will be more emphasis on discussions and a need for process owners to be available so that they can explain the processes. Open questions have always been part of assessment and there will be greater emphasis on open discussions with the organizations following the move away from documented procedure.

An important feature of QMS assessment is the ability to follow the line of sight linking the strategic direction and leadership from senior management throughout the organization. When auditing various processes the organization may not have a traditional document or even a process flow diagram. However, there will be a process owner, process objectives and interactions with other processes in the system. Using this information to review the effectiveness of the process to meet the intended outcome of the system is the key. The major challenges in ISO 9001:2015 audit are auditing context, leadership, commitment and risk.

6.8.1 Auditing Context

ISO 9001:2015 standard requires an organization to determine its context, including the needs and expectations of relevant interested parties and external and internal issues. To do this, an organization can use various techniques for strategic analysis and planning.

Auditors should confirm that suitable processes have been developed and are used effectively, so that their results provide a reliable basis for determining the scope and the development of the management system. To do this, auditors, should consider objective evidence related to following:

a. The process(es) or method(s) used.
b. The suitability and competence of the individuals contributing to this process(es).
c. The results of the processe(es).
d. The application of the results to determine management system scope and development.
e. Periodic reviews of context, as appropriate.

Auditors should have relevant sector-specific knowledge and understanding of the management tools that organization can use in order to make a judgement regarding the effectiveness of processes used to determine context.

6.8.2 Auditing Leadership and Commitment

ISO 9001:2015 standard have increased requirements for top management. These requirements include demonstrating commitment and leadership by taking accountability for the effectiveness of the management system and fulfilling a number of responsibilities. These include tasks that top management should undertake itself and others that can be delegated.

Auditors should obtain objective evidence of the degree to which top management is involved in decision-making related to the management system and how it demonstrates commitment to ensuring its effectiveness. This can be achieved by reviewing the results from relevant processes (for example, policies, objectives, available resources, communications from top management) and by interviewing staff to determine the degree of top management engagement.

Auditors should also aim to interview top management to confirm that they have an adequate understanding of the discipline-specific issues relevant to their management system, together with the context their organization operates within, so that they can ensure that the management system achieve its intended results.

Auditors should not only focus on leadership at the top management level but should also audit leadership and commitment at other levels of management, as appropriate.

6.8.3 Auditing Planning/Auditing Risks and Opportunities

Planning has always been an element of ISO 9001, but now there is increased focus on ensuring that it is considered with the context of the organization and interested parties. There is requirement to identify risks and opportunities, the impact these may have on conformity of products and services and customer satisfaction and how you plan to address these.

Although there is increased focus on risk-based thinking there is no mandatory requirement for formal methods for risk management or a documented risk management process.

As part of the assignment of an individual audit the determination and management of the organization's risk and opportunities can be included. The core objectives for such an audit assignment are to:

- Give assurance on the credibility of the risk and opportunity identification process(es).
- Give assurance that risks and opportunities are suitably determined and managed.
- Review how the organization addresses its determined risks and opportunities.

An audit of an organization's approach to the determination of risks and opportunities should not be performed as a stand-alone activity. It should be implicit during entire audit of a management system, including when interviewing top management. An auditor should act in accordance with the following steps and collect objective evidence as follows:

a. Inputs used by the organization for determining its risks and opportunities, which may include:

- Analysis of external and internal issues;
- The strategic direction of the organization;
- Interested parties, related to its discipline-specific management system and their requirements, also:
- Potential source of risk such as environmental aspects, and safety hazards, etc.

b. Method by which risks and opportunities are evaluated, which can differ between disciplines and sectors.

The organization's treatment of its risk and opportunities, including the level of risk it wishes to accept and how it is controlled, will require the application of professional judgement by the auditor.

6.8.4 Auditing Support

Support ensures that you have the right resources, people and infrastructure required to meet your organizational goals. Organizational knowledge is a new requirement in addition to competence, awareness, and communication of the QMS.

It is important for auditor to look for evidence on how to determine and provide the resources needed for the quality management system including external providers.

In relation to the competence, there has been a shift from people in the organization to anyone under the organizations control who can not only affect the product and service but can affect the performance of the QMS. The auditor will have wider sampling pool from which the objective evidence is selected.

In addition, one has to consider changing needs and trends and how one will maintain existing knowledge and acquire additional knowledge. The auditor needs to understand how to determine the knowledge needed by the organization and how it is protected.

There is less documentation required in 2015 version of the standard. There is no mandatory requirement for a quality manual or documenting the procedures for control of documents. However, there is need to ensure there are documents and records one need to ensure that the system is effective.

Auditor should know how documents and records are needed and established and verify throughout the organization to see that they are available and that they provide confidence that the processes have been carried out as planned and to demonstrate that the product or service meets the customer requirements and any regulatory or statutory requirements.

6.8.5 Auditing Operations

This part of standard covers the execution of plans and processes from initial customer interaction to the delivery of their products and services. This is linked with the actions to address risks and opportunities and control of externally provided products, processes and services and introduces a sub-clause to address changes. The auditor needs to verify how the actions which have been identified to address risks and opportunities have been implemented and controlled. Auditor needs to audit change activities and then follow this throughout the organization looking at the requirements of leadership, risk, communication, awareness, resources, competence, organizational knowledge and evaluation of performance to test its effectiveness.

The audit of supply chain to specific requirements can be required. The supplier audit programme should be developed with applicable audit criteria for the types of suppliers and external providers. The scope of supply chain audit can differ, e.g. complete management system audit, single process audit, product audit.

6.8.6 Auditing Performance Evaluation

Requirements for monitoring, measurement, analysis and evaluation are covered and we will need to consider what needs to be measured, the methods employed, when data should be analysed and when it should be reported on. The standard places more emphasis on the output of the monitoring and measuring activity.

Auditor must review the data established to achieve intended outcomes of the standard in relation to customer satisfaction and delivering products and services that meet customer statutory and regulatory requirements.

6.8.7 Auditing Improvement

The corrective action has various stages, i.e. correction, investigation and corrective action. Auditor can verify the effectiveness of corrective action.

6.8.8 Summary

The ISO 9000 quality system standards emphasize the importance of quality audits. By using the audit process, management can measure the suitability, effectiveness, and continuous improvement of its own quality system as well as the capacity of existing and prospective sub-contractors support the organization. The quality system audit also provides objective evidence concerning the need for reduction, elimination and prevention of non-conformity within the organisation. The audit program of an organization is essential to quality system measurement and continuous improvement. It inherently involves auditors from various functions of the organization which creates an ownership and commitment to the well-being of the quality system.

BIBLIOGRAPHY

1. https://committee.iso.org/home/tc176/iso-9001-auditing-practices-group-html
2. https://advisera/9001academy/why-is-iso9001
3. https://www.9001 simplified.com/learn/how-about-iso-9001-audit.php
4. ISO 19011:2018 Guidelines for Auditing Management Systems.
5. ISO 9000:2015 Quality Management System—Fundamental and Vocabulary.
6. ISO 9001:2015 Quality Management System—Requirements.
7. www.iaf.nu

7

Construction Project Management Practices and Quality Aspects

OVERVIEW

Construction project life cycle consists of project formulation and appraisal, project development, planning for construction, tender action, construction, commissioning and handing over. A construction project is an endeavour of the project team on behalf of owner/client to create a built facility as per defined functional objectives. Quality management in construction aims to achieve required functional and physical characteristics of a constructed facility through meticulous planning and effective quality management practices. Quality management concepts also have positive effect on time and cost of the project. The vital role of quality management is to ensure that a construction work is able to achieve its desired life span with least maintenance costs.

Quality planning refers to the identification of relevant quality standards and envisages that the design and specifications comprehensively incorporate the quality requirements of users and other stakeholders. Quality assurance activities include the consistent evaluation of the project performance to provide confidence that the project satisfies the relevant quality standards. Quality systems monitor project test results pertaining to the quality standards and help in identifying means to eliminate the non-conformities. The most common processes considered relevant to monitoring and measurement of quality are inspection and testing of on-site receipt of materials against specified standards and in-process inspection and testing to ensure that processes have been carried out conforming to specified requirements. Quality plans are a way of relating to specific requirements to work methods and practices to ensure that requirements will be met and processes controlled.

7.1 CONSTRUCTION PROJECT MANAGEMENT: LIFE CYCLE STAGES

7.1.1 A construction project consists of a unique set of processes consisting of coordinated controlled activities with start and end dates, performed to achieve project objectives. Achievement of the project objectives requires the provision of deliverables conforming to specific requirements. Generally in a construction project, besides the owner/client, the project manager, consultants, construction agencies and the users are the key stakeholders.

7.1.2 The different types of project delivery models in construction project are: (a) Traditional design, bid and build; (b) design, bid with variants; (c) turnkey, and (d) build, operate and transfer and its variants. Each of the delivery models can adopt different type of contracts depending upon the suitability of the contract type in relation to the nature of project, objectives and other project specific considerations. Project delivery model determines the manner in which the project is planned, designed and executed and contract administration carried out. It also determines the contractual relationships between the owner/client, design consultants and construction agency. The delivery model shall define the space of control and role and responsibilities of each of the above parties.

7.1.3 Construction project differences may occur in the deliverables through, influence of interested parties; resources used; constraints; and the way processes are tailored to provide the deliverables. Construction projects are usually organized into phases that are determined by governance and control needs. Project phases divide the project life cycle into manageable sets of activities. These phases should follow a logical sequence, with a start and an end, and should use required resources to provide desired results. In order to manage the project efficiently during the entire project life cycle, a set of activities should be performed in each phase of implementation.

7.1.4 The construction project from commencement to completion involves following stages:

    a. Project formulation and appraisal including inception, feasibility and strategic planning.

    b. Pre-construction stage including project development, planning for construction, tender action and award of work.

    c. Construction stage action including site management, construction strategy and sequence of construction.

    d. Commissioning and handing over stage including contractual closeout, financial closeout, defect liability management, facility handover, operation and maintenance.

7.2 PROJECT FORMULATION AND APPRAISAL STAGE

7.2.1 Project Formulation is systematic development of an idea, a need or an opportunity into a project proposal, outlining the project scope and objectives of the proposed project to be completed within the requirements of time, cost, quality, safety, etc. It is necessary to frame the outline of the proposed project, including scope of work involved, its duration, the preliminary cost involved, quality aspects as well as viability of the project. Construction project formulation deals with assessing scope and requirements of the project meeting the client's needs, while meeting with environmental impact assessment and sustainability aspects.

7.2.2 Every project has certain stages during its formulation before it is approved for implementation. These stages provide a framework for clear understanding of resource allocation, scheduling project milestones for implementation, and subsequently establishing a monitoring system. Various stages during project formulation and associated activities are as follows:

a. **Need for the project:** This stage involves inception of idea, and establishment of project need through "need analysis" keeping in view the existing facilities, demands and development goals.

b. **Project identification:** This is the stage of the project formulation when an idea is evaluated and potential project proposals are identified. The purpose is to establish the basic objectives of the project and identify the priorities of the project.

c. **Scope preparation:** The project scope establishes the type of works that need to be completed to deliver a product, service or a result with specified features and functions. It is the boundary of a project within which all activities in the project are required to be executed. The scope needs to be defined in measurable parameters, without any ambiguity to avoid subsequent deviations and unnecessary work. It could accommodate minor variations that may arise during project execution. The purpose of scope preparation of proposed work is to achieve clarity of the project scope, including its objectives and deliverables.

d. **Availability of land:** Land is the single most important factor, without availability of which, the process of project formulation will be incomplete. A detailed survey of land has to be carried out before acquiring land and related aspects pertaining to land, need to be investigated and clearly brought out in draft project formulation report.

e. **Prefeasibility study:** This is the first attempt to examine the overall potential and viability of the project, when all preliminary steps for going into a detailed feasibility are exercised. A conscious decision needs to be taken after prefeasibility study whether next stage of feasibility study should be ventured into.

f. **Feasibility study:** The project details are examined to see, if they have promise of meeting the financial, economic, environmental, sustainability and social criteria that has been set for investment expenditure. Engineering investigations and detailed studies are carried out at this stage to assess if the project is feasible or not. The feasibility report will form the basis for whether to go in for detailed proposal or not, by evaluating internal and external constraints and other parameters.

g. **Financing and cost benefit analysis:** It involves estimating the project costs including cost of capital, opportunity cost, operating cost and fund requirements for comparing various project proposals on a common scale. It also provides a platform for investment decisions involving commitment of resources in future, with a long term horizon. Cost benefit analysis gives idea of benefits of the project and provides platform for preparation of detailed project report. The overall worth of project is established by evaluating the cost benefit analysis output connected with the project.

h. **Pre-investment analysis:** The results obtained in previous stages are consolidated to arrive at clear cut conclusions. It helps the project sponsoring body, the project implementation body and the external consulting agencies to accept or reject the proposal.

7.2.3 The work of project formulation shall culminate in preliminary project formulation report, which should broadly include the following: (a) Objectives of the project, (b) salient features of the project, (c) cost estimates and source of funds, (d) preliminary project cost, (e) land for the project, (f) feasibility report, (g) preliminary quality and safety assurance plan, (h) procurement strategy, (i) project program sequencing, (j) mobilization and staffing requirement, (k) statutory approvals required, (l) economic and financial analysis, and (m) conclusion and recommendations.

7.2.4 Project appraisal it is the process of critically assessing and reviewing project proposals by validating the assumptions, conducting the necessary technical, physical and financial feasibility studies before decision to implement the project is taken. It is necessary for a review whether the basic assumptions have sound grounds and proposal meets with intended objectives and requirements; the cost indicated is reasonable and can be financed with the resources available; and the aspects of time, quality, sustainability; and other items can be taken care of. Project appraisal is done by the owner, generally through specialized agencies, well experienced in such work, appointed by owner or funding agencies. It may be necessary to have separate team within the organization or work may be entrusted to an experienced consultant in the field. If necessary, work of project appraisal may be got done through combination of in house expertise and consultants.

7.3 PRE-CONSTRUCTION STAGE

7.3.1 Project Development

Project development shall include: Site survey and soil investigation; alternative concept designs with costing development of design of each discipline and their integration; obtaining statutory approvals; decision on construction methodology; detailed design of each discipline; construction drawings and related specifications with integration of engineering inputs of all concerned disciplines; detailed cost estimates; detailed specifications and bill of quantities; including tender documents.

Peer review/proof checking of the drawings/designs/estimates shall be done in case of important projects, depending upon their complexity and sensitivity. Environment impact analysis and social impact analysis shall be done in applicable cases as required.

7.3.2 Planning for Construction

Methodology of construction shall be detailed before the start of the project. Sequencing of project components shall be done on the basis of methodology adopted and availability of resources. This shall be reviewed during the progress of the project and revised, if necessary. The planning tools like Work Breakdown Structure (WBS), bar chart, network technique and scheduling shall be employed for effective management of a construction project.

Resource planning shall involve the following: (a) Resource allocation—the feasibility of the network shall be checked with respect to manpower, equipment, other resources required at the site. (b) Resource levelling—it shall be done by reallocating the slack resources from non-critical path to critical path activity in order to obtain a reduction of time or by shifting the activities within the floats available with them, to obtain an optimum uniform resource requirement. (c) Resource schedule—schedule of resource requirements with respect to time shall be prepared on the basis of network developed and kept in the database for project control purposes in respect of: (1) Technology, (2) manpower: (i) technical staff, (ii) skilled labour, and (iii) unskilled labour, (3) machinery, (4) materials, (5) cash flow.

Resource schedule shall be prepared separately for client, consultant and construction agency. Time cost trade-off analysis shall be done to obtain a minimum total cost of the project within the specified time. This shall be done taking into consideration direct costs and indirect costs of the project.

Construction related organizations generally establish strategy based on their mission, vision, policies and factors affecting implementation of the project. Construction projects are the opportunities which help to accomplish strategic goals. The goals and benefits may result in a justification for the investment in the project. The purpose of the justification is usually to obtain organizational commitment and approval for such an investment. Planning for the establishment and maintenance of a quality management system is a strategic process. This planning should be performed by the project organization so as to focus on the quality of processes, products and services.

7.3.3 Tender Action and Award of Work

Preparation of tender documents has to be carried out on the basis of bill of quantities, specifications, drawings and conditions of contract relevant to the project. In case of any special feature in the project implementation, the same shall be described clearly to avoid any ambiguity. Selection of contractor shall be done through open competitive bidding where tender publicity should be adequate to obtain competitive tenders. In large, specialized and important works, prequalification of contractors shall be done considering the financial capacity, experience of similar type of works, past performance, required technical staff and plants and machinery. After receipt of tenders and due evaluation and negotiation with the bidders, the work shall be awarded to the construction agency based on competitive, technical and financial bids.

7.4 CONSTRUCTION STAGE

7.4.1 General

Construction is the most important stage of construction management where pre-construction stage outputs are realized into physical tangible form within the constraints of time and cost. The functional and physical characteristics, defined in the pre-construction stage outputs through specifications, drawings and consolidated project requirements have to be realized through various construction project management functions as detailed below:

a. Procurement management includes processes for purchase of materials, equipment, product, etc. One of the fundamental issues in construction projects is to determine what needs may be met by procuring product, services and works from external agencies and what should be accomplished by the project team.

b. Time management aims to complete the project within the stipulated time period. This includes activity duration estimating, activity sequencing with inter activity dependent functions, project schedule development and project schedule control.

c. Cost management is to ensure that project is completed within the authorised budget through proper resource planning, cost budgeting and cost monitoring and control.

d. Quality management in construction aims to achieve quality objective of customer's requirements through performance evaluation of construction processes, and ensuring that these are directed towards overall quality. It has to be ensured that design, drawings and specifications comprehensively incorporate requirements of users and other stakeholders.

e. Risk management determines actions required for reducing impact of risk. Risk responses are established and assigned to appropriate project participants. Suitable risk mitigation measures should be evolved for identified risks.

f. Communication management aims to have a proper management information system (MIS) so that at the construction stage of the project, there is appropriate transmission of required information at the level of various agencies like client, architect, engineer, project manager, consultants, material suppliers, construction agencies and sub-contractors.

g. Human Resource Management aims to have human resources in project which are adequately trained and competent. A formal training or certified course undertaken should be a preferred selection criterion for the workers. A periodic review of the performance may be made to establish the nature of training required and methods for imparting additional training.

h. Health and safety management issues include looking into the risk factors to health of construction personnel and providing hygienic conditions at construction sites and methods of their management. Safety management issues include managing work processes, equipment and material handling at site and striving to achieve zero accident status at site.

i. Sustainability management issues include minimizing adverse environmental impact of activities, products and services. It also includes limiting the adverse impact within the laws and prescribed norms and their monitoring. Management of disposal of construction waste from the sites is another important issue.

7.4.2 Site Management

The site management needs to be carried out through suitable site organization structure of client, consultant and contractor with roles and responsibilities assigned to construction personnel for various project related functions. The layout of the construction site should be planned keeping in view the various requirements of construction activities and specific features pertaining to size shape, topography, etc. The site layout needs to take into consideration the following factors:

a. Easy access and exit to site, with proper parking and movement spaces for vehicles and equipment, adequate yard lighting and lighting for night shift.

b. Adequate stack areas for construction material stores and stack areas for bulk construction materials.

c. Temporary buildings, site office, shelter and sanitation facilities for workers and site laboratory.

d. Fabrication yards for reinforcement assembly, concrete pre-casting and shuttering materials.

e. Fencing, barricades and signages.

f. Access for fire fighting equipment and effective drainage plan

g. Temporary and permanent installation for construction (water, power, hoists, cranes, elevators, batching plants, etc.)

7.4.3 Construction Strategy and Construction Sequence

Construction strategy and construction methods need to be evolved at planning and design stage and implemented by the site personnel to ensure ease of construction and proper integration of construction activities. The construction sequence should be followed taking into consideration the following:

a. Availability of resources (men, material and equipment).

b. Construction methods employed including prefabrication.

c. Design requirements and load transfer mechanism.

d. Stability of ground like in hilly terrain.

e. Ensuring slope stability with retaining structure before the main construction.

f. Installation and movement of heavy equipment like cranes and piling equipment.

g. Effect of weather

h. Minimum time to be spent on working below ground level.

7.5 COMMISSIONING AND HANDING OVER STAGE

7.5.1 Commissioning and Handover of Project

When all the construction related activities of the project get completed as per designs drawings and specifications, it shall further need attending to the following aspects:

a. Clearing of site and restoration of surroundings.

b. Removal of all defects at the time of completion and during defect liability period.

c. Preparation of list of inventories.

d. Finalization of construction agency's final bills.

e. Obtaining completion certificate from local government bodies and departments.

f. Preparation of maintenance manual.

g. Preparation and handing over of all other relevant documents to client including as built drawings.

7.5.2 The commissioning stage helps in identifying any remaining site activities that need to be coordinated. It helps in bringing to attention various common problems encountered. These and other problems should be identified and corrected during the commissioning process. Required resources should be allocated to commission the systems after they are installed to ensure that they work as intended. A complete and thorough functioning of the structure shall be carried out to ensure that the systems work as required.

7.5.3 The first step in the commissioning process should involve formation of a commissioning team that comprises the owner, users, occupants, operation and maintenance (O&M) staff, and design professionals. The project design documents should include a commissioning plan. Prior to the handover stage, the commissioning team should verify the installation of the systems, conduct functional performance testing, training of the O&M personnel, etc.

7.5.4 Operation and Maintenance

O&M programmes that focus on improving energy efficiency of building systems can help save energy without a significant capital investment. From small to large sites, these savings can represent significant savings each year, and can be achieved with minimal cash outlays. Effective operation and maintenance is one of the most cost effective methods for ensuring reliability, safety and efficiency. Good maintenance practices result in substantial savings in consumption of energy and water, and should be considered as a resource.

7.5.5 Performance Tracking

Subsequent to the commissioning and handover stage, regular monitoring of the performance shall be carried out which will provide information on whether the set environmental performance and targets have been met or not.

7.6 ISSUES RELATED TO PROJECT MANAGEMENT

7.6.1 Significance of Project Management in Construction Projects

A construction project is generally faced with challenges of uncertainties leading to time-over runs, cost over-runs, changes in project parameters, loss of quality, and inability to meet functional objectives. All such factors underpin the need of effective and efficient project management in construction projects. Project management is application of methods, tools, techniques and competencies to a project which facilitates project to be completed within scheduled time, cost and to the requirement of quality standards. Project management includes the integration of the various phases of the project life cycle.

Construction planning and the site management play an important role in smooth progress of construction activity. The knowledge of actual technical provisions in regard to practices relating to various components of construction work starting from sub-structure to super structure play a key role in achieving requisite quality, durability and finish.

Besides technical aspects, management aspects also play an important role in the success of project. Therefore, management functions and technical processes in a construction project need to be integrated towards achieving project objectives.

7.6.2 Project Constraints

There are several types of constraints which are often interdependent. The project deliverables should fulfil the requirements for the project and relate to any given constraints such as scope, quality, schedule, resources and cost. Constraints are generally interrelated such that a change in one may affect one or more of the other constraints. Hence, the constraints may have an impact on the decisions made within the project management processes. Achievement of consensus among key project stakeholders on the constraints may form a strong foundation for project success.

Some of the constraints requiring attention are as follows:

a. The duration or target date for the project.

b. The availability of the project budget.

c. The availability of project resources, such as people, facilities, equipment, materials, infrastructure, tools and other resources.

d. Factors related to health and safety of personnel.

e. The level of acceptable risk exposure.

f. The potential social or ecological impact of the project.

g. Laws, rules and other legislative requirements.

7.6.3 Project Governance

Governance is the framework by which an organization is directed and controlled. Project governance includes, but is not limited to, those areas of organizational governance that are specifically related to project activities. Project governance may include subjects such as the defining the management structure; the policies, processes and methodologies to be used; limits of authority for decision-making; stakeholder responsibilities and accountabilities; interactions such as reporting and the escalation of issues or risks.

Project governance may involve the following:

a. The project sponsor, who authorizes the project, makes executive decisions and solves problems and conflicts beyond the project manager's authority.

b. The project steering committee or board, which contributes to the project by providing senior level guidance to the project.

c. Customers or customer representatives, who contribute to the project by specifying project requirements and accepting the project deliverables.

d. Suppliers, who contribute to the project by supplying resources to the project;

e. The project management office, which may perform a wide variety of activities including governance, standardization, project management training, project planning and project monitoring.

7.6.4 Stakeholders and Organizations

The project stakeholders, including the project organization, should be fully involved for the project to be successful. The roles and responsibilities of stakeholders should be defined and communicated based on the organization and project goals.

The "originating organization" is the organization that decides to undertake the project. It can constitute as a single organization, joint-venture, consortium or any other acceptable structure. The originating organization can undertake multiple projects, each of which should be assigned to a different project organizations.

The "project organization" carries out the project. The project organization may be a part of originating organization. There should be clear division of responsibilities between project organization and other relevant interested parties (including the originating organization) for the project processes. These should be maintained as documented information. Stakeholder interfaces should be managed within the project through the project management processes.

The project organization may include the following roles and responsibilities:

a. The project manager, who leads and manages project activities and is accountable for project completion.

b. The project management team, which supports the project manager in leading and managing the project activities.

c. The project team, which performs project activities.

7.6.5 Competence of Project Personnel

The project team requires competent individuals who are capable of applying their knowledge and experience to achieve the project deliverables. Any identified gap between the available and required competence levels in the project team could introduce risk, and needs to be addressed. Competency levels may be raised through professional development processes such as training and mentoring inside or outside the organization.

7.6.6 Project Management Processes

Project management includes the planning, organizing, monitoring, controlling and reporting of all processes of a project, taking the necessary corrective and improvement actions, that are needed to achieve the project objectives, on a continual basis. The quality management principles should be applied to all project implementation processes.

The processes that are applicable to any project phase or project can be identified/defined through following aspects: Initiating; planning; implementing; controlling and closing processes as detailed below:

a. The initiating processes are used to start a project phase or project, to define the project phase or project objectives and to authorize the project manager to proceed with the project work.

b. The planning processes are used to develop planning detail. This detail should be sufficient to establish baselines against which project implementation can be managed and project performance can be measured and controlled.

c. The implementing processes are used to perform the project management activities and to support the provision of the project's deliverables in accordance with the project plans.

d. The controlling processes are used to monitor, measure and control project performance against the project plan. Consequently, preventive and corrective actions may be taken and change requests made, when necessary, in order to achieve project objectives.

e. The closing processes are used to formally establish that the project phase or project is finished, and to provide lessons learned to be considered and implemented as necessary.

7.6.7 Quality Management Processes

It is necessary to manage project processes along with a quality management system in order to achieve project objectives. Where the project organization operates within the originating organization, the project quality management system should be aligned as far as possible, with the quality management system of the originating organization. Where a part or all the project organization is external to the performing organization, quality management system requirements may need to be specified to ensure that project processes are capable of proper interfacing. Documented information needed and produced by the project organization to control of the project should be defined and implemented.

7.7 QUALITY PLAN

7.7.1 General

A quality plan describes how an organization will provide an intended output, whether that output is a process, service, project or contract. The quality plan should state the quality objectives for the specific construction/design project and how they will be achieved. Quality objectives may be established, for example, in relation to: Quality characteristics for the specific construction/design project; important issues for satisfaction of the customer, organization or other interested parties including opportunities for improvement. These quality objectives should be expressed in measurable terms. Any required measurement processes needed to determine achievement of the quality objectives should be included or referenced in the quality plan.

Principal employer may choose to request contractor/consulting organization to submit a quality plan. Both the principal employer organization requesting a quality plan and contractor/consulting organization should consider the reasons for using a quality plan and the benefits that might be achieved through its use. The principal employer/organization requesting a contractor/consulting organization quality plan should apply risk-based thinking to the nature of the design/construction work, the evaluation and selection of external provider(s) and opportunities for benefits. There can be benefits to both the organization and potential external providers in using proper quality plan.

Principal employer requirements for contractor/consulting organization quality plans can be included in specifications for other management plans such as service management plans, project management plans, construction management plans or production and installation plans. A quality plan can ensure that a principal employer of the organization has a common understanding with contractor/consulting organization about how its requirements will be met. The principal employer organization should decide what level of monitoring is required to assess contractor's/consulting organization's performance, such as ongoing monitoring, acceptance checks, assessment and auditing.

Establishing a common understanding of the quality plan between the principal employer organization and the contractor/consulting organization is particularly important where the specific case involves high levels of risk and complexity. A common understanding means that the organization has a basis for confidence in satisfactory performance by the contractor/consulting organization and the contractor/consulting organization has a basis for communicating with the principal employer organization about potential problems.

7.7.2 Process Approach in Quality Plan

The process approach means the systematic management of processes and their interactions to achieve intended results. Applying the process approach to quality plans assists organizations to manage the inputs, activities and outputs of each process within a coherent system of interrelated processes. Processes referred in the quality plan can interact with: Each other (interactions among quality plan processes); other processes operated within the organization's management system; processes operated within other organizations (such as customers and external providers).

7.7.3 Development of a Quality Plan

Understanding the context of the quality plan and its intended results provide a basis for determining risks and opportunities to be addressed. The context of the quality plan can include: Existing management plans or processes which will support the quality plan, whether or not these processes are part of an established management system; internal issues that can affect the ability of the contractor/consulting organization to achieve the intended results, such as constraints on resources, how the quality plan will be communicated to its users; external issues related to the specific case, such as statutory and regulatory requirements, competitive and market issues; the aspects of both the internal and external issues of the organization that relate to the specific construction/design work, for example, quality objectives; the needs and expectations of relevant interested parties, including principal employers, employees of construction/consulting organization, external providers, etc.

Risks should be determined and addressed, in order to provide confidence that intended results will be achieved and undesired effects will be prevented or reduced. Opportunities for improvement should be considered, for example, to meet customer expectations or increase effectiveness and efficiency. Opportunities for innovation can also be important, for example, where draft quality plans are submitted as part of a tendering process for provision of products and services.

The contractor/consulting organization should determine what is to be covered by the quality plan. The scope of the quality plan will depend on several factors, including: The requirements of customers and other relevant interested parties; the types of products and services to be provided; the organization's processes and their quality characteristics; the resources needed to achieve the intended results; the extent to which the quality plan is supported by an established quality management system. There can be benefits from reviewing the scope of the quality plan with the principal employer or other relevant interested parties. In preparing the quality plan, the contractor/consulting organization should determine the respective roles, responsibilities and authorities within the organization and, where applicable, the relevant responsibilities and authorities of external parties such as principal employer.

The quality plan should identify people within the organization who are responsible for: Ensuring that the activities and resources required for the quality plan or contract are planned, implemented and controlled, and their progress monitored; reviewing quality plan inputs, recording these reviews and resolving conflicts and ambiguities; communicating requirements to all affected departments and functions, external providers and customers/principal employer, and resolving problems that arise at the interfaces between such groups; reviewing the results of any audits conducted; reviewing and authorizing changes to, or deviations from, the quality plan.

7.7.4 Operation and Control of Quality Plan

The quality plan should be reviewed for adequacy and effectiveness, and should be formally approved by an authorized person or a group that includes representatives from relevant functions within the contractor/consulting organization. In contractual situations, a quality plan might need to be submitted to the principal employer/customer by the organization for review and acceptance, either as part of a pre-contract consultation process or after a contract has been awarded. Once a contract is awarded, the quality plan should be reviewed and, where appropriate, revised to reflect any changes in requirements.

7.7.5 Implementation of Quality Plan

In the implementation and monitoring of the quality plan, the contractor/ consulting organization should consider the following issues: Distribution of the quality plan to all relevant people; the control provisions for documented information; training in the use of quality plans; special training could be needed to assist users in applying the quality plan correctly monitoring in conformity with quality plans; the organization is responsible for monitoring conformity with each quality plan that it operates, which can include: Operational supervision of the planned arrangements; milestone reviews; audits, etc.

Whether carried out by internal or external interested parties, such monitoring can assist in: Assessing the commitment of the organization to the effective implementation of the quality plan; evaluating the practical implementation of the quality plan; determining where risks can arise in relation to the requirements of the specific construction/design works; taking corrective action where appropriate; finding opportunities for improvement in the quality plan and associated activities.

7.7.6 Revision and Improvement in Quality Plan

The organization should revise the quality plan: To reflect any changes to quality plan inputs or risks, including: the specific case for which the quality plan is established; the processes for production and service provision; the organization's management system; statutory or regulatory requirements; to incorporate agreed improvements to the quality plan.

An authorized person or persons should review changes to the quality plan for impact, adequacy and effectiveness. Revisions to the quality plan should be made known to users, customers, interested parties and/or external providers. Communication with customers and other interested parties should be consistent with the requirements for externally provided products and services. Any documented information that is affected by changes in the quality plan should be revised as necessary.

7.7.7 Cost of Quality

Cost of quality is defined as a methodology that allows an organization to determine the extent to which its resources are used for activities that prevent poor quality, that apprise the quality of organization's product and services, and that result from internal and external failures. In simple way we can say that cost of quality comprises cost of conforming quality and cost of not conforming quality. The cost of conforming quality is categorized into prevention cost and appraisal cost, whereas cost of not conforming quality categorized into cost of internal failures and cost of external failure. Cost of quality is addition of all four categories of cost. Quality cost analysis is useful for setting priority for top management. Trend analysis of cost of quality helps in monitoring the progress of improvement actions. It is often seen that in critical stages, share of failure costs in total cost of quality is predominant.

The prevention costs are the costs of such activities undertaken to prevent defect in design, development, purchase, labour, and other associated items involved in creation of products and services. Prevention is achieved by examining the previous failure data and developing action plans for incorporating the same into basic system so that failure and defects do not occur again.

Appraisal costs are the costs incurred which conducting inspections, tests and other planned evaluation with the purpose of determining whether the product or service confirms to its stated requirements. Appraisal cost also includes various activities related to quality system audit, cost of legal compliance, product quality audits, cost of calibration of testing equipment, etc.

Internal failure costs are the costs which are associated with the defective or non-conforming works. Whenever the quality appraisals are carried out, there exists possibility of discovering non-conforming situations. Such situations are generally salvaged by rework, complete replacement or scrapping. The cost of carrying out re-inspections, re-tests, failure analysis, evaluation, disposition and subsequent actions are the internal failure costs. In short, this includes all material, labour, energy and overhead expenses that are wasted on accounts of non-conforming or defective products and service.

External failure costs pertain to defects which are found after the product reaches the customer. This component of cost also disappears if there are no defects. Examples of the external failure cost include complaints from customers, cost involved in repairs or replacement of products during warranty period. Retrofit or recall cost may be involved in case of requirement of modify or update the product to overcome the design deficiencies. Insurance claims and contractually obligatory claims also included in such type of costs. External failure will also include loss of goodwill leading to loss of sales.

7.8 QUALITY MANAGEMENT IN PROJECTS

7.8.1 General

The project quality management system should be documented or referenced in the project's quality plan. Linkage should be established between the project's quality plan and applicable parts of the quality management system. The project organization should adopt the quality management system and processes of the originating organization. In case where, specific requirements for the quality management system from other interested parties exist, it should be ensured that the project's quality management system is compatible with these requirements.

The project organizational structure should be established in accordance with the requirements and policies of the originating organization and the conditions particular to the project. The project manager and the originating organization should ensure that the project organization structure is appropriate to the project scope, the size of project team, local conditions and the processes employed. The project function responsible for ensuring that the project's quality management system should be established implemented and maintained. The interfaces of this function with other project functions, the customer and other interested parties should be documented.

Both the project organization and the originating organization should consider the context in which their project quality management systems operate. Some internal and external issues can affect the project's ability to achieve the intended project results. Other issues can offer opportunities to work more effectively with internal and external parties.

7.8.2 Management Responsibility in Projects

The commitment and active involvement of the top management of both the originating and project organization are essential for the developing and maintaining an effective and efficient quality management system for the project. Top management of both the originating and project organizations should create a culture for quality which is an important factor in ensuring the success of project and could provide input into the strategic processes planned to be performed by the project organization.

The project organization's managers responsible for the project should review project's quality management system, at planned intervals, to ensure its continuing suitability, adequacy, effectiveness and efficiency. Progress evaluations should cover all the project's processes and provide an opportunity to assess the achievement of the project objectives. The outputs of the evaluation should be assessed against the project objectives, to determine whether the performance of project against the planned objectives is acceptable. The outputs from progress evaluation can provide significant information on the performance of the project as an input into future management reviews.

7.8.3 Process Control and Non-conforming Products

The organization should identify and plan the production installation and processes. That directly affects the quality and shall ensure that these are carried out under controlled conditions. Documented procedures defining manner of production, use of equipment and machinery, compliances with reference to standards and codes, monitoring control and workmanship constitute important components of process control.

The organization shall establish and maintain documented procedures for carrying out inspections and testing of activities in order to verify that specified requirements for the products are met with. Incoming product should not be used prior to inspection and testing. In progress inspection and testing shall be carried out as required by quality plan including documented procedure to ensure conformance to specifications of the finished product. Documented procedures to control, calibrate, maintain inspections and measuring and testing requirements shall be established. The equipment should be, kept in sound operating condition, calibrated at specified frequency and used as per instructions.

In respect of non-conforming products, that do not conform to product requirements, shall be identified and controlled to prevent its unintended use on delivery. Proper procedures shall be established to define the controls and related responsibilities and authorities to deal with non-conforming products. Where applicable, the organization shall deal with non-conforming products by one or more of the following ways:

a. By taking action to eliminate the detected non conformity.
b. By authorizing its use, release or acceptance under concession by relevant authority.
c. By taking action to preclude its original intended use for specific application.
d. By taking action appropriate to the effects of non-conformity, when non-conformity of product is detected after delivery.

When non-conforming product is suitably modified or corrected, it shall be subject to re-verification to determine conformity to the requirements. Records of the nature of non-conformities and any subsequent action taken, including concession obtained, shall be maintained.

Control of non-conforming products is important to ensure that such products are prevented from inadvertent use of installation. Non-conforming products shall be reviewed in accordance with documented procedures for, (a) reworking to meet the specifications, (b) accepted with or without repair with concession, (c) re-graded for alternative use, or (d) rejected or scrapped.

Improvement through use of quality policy, quality objectives, audit results, analysis of data is required to continuously improve the effectiveness of quality management system.

Corrective action is required to eliminate the causes of non-conformities in order to prevent the recurrence. Corrective action shall be appropriate to the effects of the non-conformities encountered. A documented procedure shall be established to deal with requirement for, (a) reviewing non-conformities including customer complaints, (b) determining causes of non-conformities, (c) evaluating the need to work out methodology to ensure that non-conformities do not occur, (d) determining and implementing required action, and (e) reviewing the effectiveness of the corrective action taken.

Prevention action is required to eliminate the causes of potential non-conformities in order to prevent their occurrence. Preventive action shall be appropriate to the effects of potential problem. A documented procedure shall be established to determine requirements for, (a) determining potential non-conformities and their causes, (b) evaluating the need for action to prevent occurrence of non-conformities, (c) reviewing the effectiveness of the preventive action taken.

7.8.4 Resource Management in Projects

Resources needed for the project should be identified. The resource related processes aim to plan and control resources. Examples of resources include people, equipment, facilities, finance, information, materials, computer software and services.

Resource plans should state what resources will be needed by the project and when they will be required, according to the project schedule. The plans should indicate how, and from where, resources will be obtained and allocated. The plans should also include the manner of disposition of resources at the completion of tasks and at the end of project. Reviews should be performed to ensure that sufficient resources are available to meet the project objectives.

7.8.5 Measurement Analysis

Both originating and project organizations should learn from the project processes and should apply methods for correcting, preventing or reducing undesired effects to enable continual improvement in both current and future projects. They should ensure that the measurement, collection and validation of data are effective and efficient, in order to improve the originating organization's performance and to enhance the satisfaction of customer and other interested parties.

Examples of measurement of performance include: (a) Evaluation of individual activities and processes; (b) auditing; (c) evaluation of actual resources used, together with cost and time, compared to the original estimates; (d) evaluation of products and services; (e) evaluation of external provider performance; (f) achievement of project objectives; (g) satisfaction of the customer and other interested parties.

The managers responsible for the project from the project organization should ensure that documented information on nonconformities and the disposition of nonconformities in the project's products/services and processes is analysed to assist learning and to provide data for improvement. The project organization, in conjunction with the customer, should decide which nonconformities should be recorded and which corrective actions undertaken.

7.8.6 Information Management in Organization

The originating organization should define the information it needs to learn from projects and establish a system for identifying, collecting, storing, updating and retrieving information from projects. The originating organization should maintain a list of all risks managed by its projects. The originating organization should ensure that relevant information is used by other projects that it originates.

The project organization should design the projects information management system to implement the requirements specified for learning from the project by the originating organization. The project organization should ensure that the information it provides to the originating organization is accurate and complete. The project organization should implement improvements using information relevant to the project, which has been established by the originating organization using the system.

7.8.7 Product and Service Realization in Projects

The organization shall plan and develop the process needed for product realization. Planning of product realization shall be consistent with the requirements of the other processes of QMS. The organization shall determine the following as appropriate, (a) quality objectives and requirements of the product, (b) the need to establish process and documents and to provide resources specific to the product, (c) required verification, validations, monitoring, inspection and test activities specific to the product and criteria for product acceptance, and (d) records needed to provide evidence that the realization processes and resulting product or service meet requirements. The output of this planning should be in a form suitable for the organization's methods of operations. A document specifying the processes of QMS including the product realization processes and the resources to be applied to a specific project or contract shall be prepared and used for implementation.

Projects consist of a system of planned and interdependent processes and an action on one of these usually affects others. The interdependent processes are: (a) Project initiation and project management; (b) plan development; (c) interaction management; (d) change management; (e) processes and (f) project closure.

To facilitate the interdependencies (which are planned) between processes, the interactions (which are not planned) in the project should be managed. This should include: Establishing processes for interference management; holding project interfunctional meetings; resolving issues, such as conflicting responsibilities or changes to risk exposure; measuring project performance; carrying out progress evaluations to assess project status and to plan for the remaining work.

Change management in project is quite important as it can impact the project performance and achievement of project objectives. Change management covers the identification, evaluation, authorization, documentation, implementation and control of change. The impact of change should be identified as soon as possible. The root causes of the impacts should be analysed and the results should be used to produce solutions and to implement improvements in the project processes.

At the closure of project there should be a formal handover of the project product and service to the customer. Project closure is not completed until the customer formally accepts the project product/service.

BIBLIOGRAPHY

1. IS 15883 (Part 1): 2009: Construction Project Management—Guidelines

2. IS 15883 (Part 4): 2015: Construction Project Management—Guidelines, Quality Management

3. IS 16416:2016 Construction Project Management: Project Formulation and Appraisal–Guidelines

4. ISO 10005:2018: Quality Management Guidelines for Quality Plan.

5. ISO 10006:2017: Guidelines for Quality Management in Projects.

6. ISO 21500: Guidance on Project Management.

7. ISO 9001:2015: Quality Management Systems—Requirements.

8

Implementation of ISO 9000 Practices in Construction— Case Studies

OVERVIEW

In today's construction scenario in our country, the quality product is no more a choice but an inevitable requirement for growth and prosperity. The entire construction industry including architects/engineers, contractors, sub-contractors, manufacturers of construction materials and products are benefitted from implementation of ISO 9000 standards in their functioning. During last decade of 20th century, it was noted that some principal employers, construction agencies and consultants adopted ISO 9000 practices in their organizations. The case studies discussed are as follows:

i. **Case Study I:** Pertains to Parliament Library Building Project in 1993–94, a multi-dimensional institution of national importance. The certificate of ISO 9002 quality system was obtained through Bureau of Indian Standards by CPWD for execution of the works. It helped in achieving the desired goal of quality consistent with quality policy for the project.

ii. **Case Study II:** Construction of flyovers in Delhi during 1997–2000, PWD Delhi obtained ISO 9002 certification for establishing and maintaining a quality system for construction of various flyovers and executed the flyover projects in a time bound manner. The Indian Roads Congress published "Guidelines on Quality Systems for Road Bridges, IRC SP-47, which was also followed to align construction activities with ISO 9002 requirements.

iii. **Case Study III:** Pertains to Ahluwalia Contracts (I) Ltd. New Delhi, a leading construction agency of India has found ISO management systems certification as an important practice to create, develop and promote an integrated approach towards core business processes, customer focused processes, achieving better productivity and continual improvement.

iv. **Case Study IV:** Intercontinental Consultants and Technocrats Pvt Ltd., New Delhi, has experienced ISO 9001 QMS certification as an important management tool responding to changes in business environment. The practice of ISO 9000 standards brings a set of procedures which define methods to be followed for doing a task, bringing in uniformity and continuity across time lines, helping to align a strategy consistent with the vision of the management.

8.1 CASE STUDY I: IMPLEMENTATION OF ISO 9000 PRACTICES IN CONSTRUCTION OF PARLIAMENT LIBRARY BUILDING PROJECT, NEW DELHI

8.1.1 Introduction

- Central Public Work Department (CPWD), the premier construction agency of Govt. of India, had taken up the execution of Parliament Library Building Project adjoining the Parliament House in Lutyens, Delhi. The Parliament library building was conceived to be a landmark in the capital equipped to meet the basic requirements of a most modern library with computer linkage with libraries all over the world, especially the UN library. It was to be developed as a multi-dimensional institution of national importance to be used by VIPs, foreign dignitaries, scholars and persons of eminence. It was considered essential to follow a quality assurance system that could provide confidence that the desired quality will be achieved in the construction and services.
- ISO-9002 quality systems, a model for quality assurance in production and installation, was considered to be relevant for certification. The Parliament Library Project (PLP) civil wing of CPWD became the first civil engineering organization in the country to be have been conferred with ISO-9002 quality systems certification by Bureau of Indian Standards (BIS).

8.1.2 Management Responsibility

8.1.2.1 *Quality Policy*

- The quality policy of the Parliament Library Project Team (hereinafter referred as PLP) had been defined as follows:
- The PLP's endeavour will be to ensure that: (a) The building work assigned to the PLP meets the functional needs of the client, (b) aesthetically the building conforms to architectural drawings and specifications for finishing items as proposed by the architect; there is no time and cost over-runs to the extent possible for which appropriate corrective action shall be taken from time to time. The policy is understood and is being implemented and maintained at all levels of the PLP.

8.1.2.2 *Organization*

- A consultant architect had been appointed for architectural design and to ensure aesthetics of the proposed building. Structural design was assigned to Central Designs Organization of the department. For planning including services, project management and execution, the project team was functioning from the project site. The civil wing of PLP was a compact team of engineers and other technical staff. It consisted of Chief Engineer and Project Manager-1 no. Superintending Engineer/Assistant Executive Engineers-6 nos, Junior Engineers-10 nos and Draughts men-2 nos.

8.1.2.3 *Responsibility and Authority*

The project team members (Junior Engineers, Assistant Engineers/Assistant Executive Engineers and Executive Engineers) were delegated well defined responsibility and authority to manage and perform their work. They were also given freedom and authority to stop and reject sub-standard work and also to take action to prevent recurrence of any non-conformity.

8.1.2.4 *Resources*

The resource requirements and budget to provide adequate resources, including the assignment of trained personnel for management, performance of work and verification activities including the internal quality audits were identified.

8.1.2.5 *Management Representative (MR)*

The superintending engineer, Parliament Library Project had been appointed as the management representative (MR), who regardless of any other duties was to be responsible for establishing, implementing, maintaining and reporting on the quality system. He was to keep the Chief Engineer and Project Manager informed of any quality problem or shortcomings in the system and recommended remedial measures and solutions and follow up the implementation.

8.1.2.6 *Management Review of Quality System*

A Management Review Committee with superintending engineer as its Chairman and the Executive Engineers as it members had been set up to look into the effective functioning of the quality system and to suggest modifications, required, if any. The review of the system was conducted periodically by committee members.

8.1.3 Quality System Documentation

The PLP team decided not to engage any outside consultant for preparation of the basic documents, viz. quality manual and procedure manual. By meticulous planning and following a systematic approach, it became possible to evolve a quality system conforming to the provisions of ISO 9002. Emphasis was given on documenting the quality system to make sure that the staff knew what they were expected to do.

8.1.4 Contract Review

The contract review was conducted to ensure Parliament Library Project team understood and was capable to meet the requirements specified by the Lok Sabha Secretariat. Meetings/interactions during the progress of work were also being held periodically and minutes of the discussions were recorded and documented. The changed or additional requirements were met under the same contract.

8.1.5 Design Control

Architectural design was being done by the consultant architect and structural design by the Central Designs Organization of the department. The design meetings were held with the client and modifications required, were made in the architectural/structural drawings.

8.1.6 Document and Data Control

A master list of various documents with up-to-date revisions concerning the quality systems requirement were maintained. Registers for various tests were issued after making an entry in the master list and their pages were numbered.

All structural and architectural drawings bearing proper identification numbers were issued through a drawings register. Test registers were periodically scrutinized by senior officers. All the documents including quality manual, procedure manual, etc. were controlled in order to prevent any unauthorized changes in them.

8.1.7 Procurement of Services

For selection of the principal architect a limited architectural companies were short-listed among the renowned Delhi based architects. The architect who was adjudged the best by the board of assessors was ultimately selected for the project.

On account of large amount of excavation particularly in rock and that too to be done without resorting to blasting, use of heavy earth moving and rock cutting machinery became inevitable. The construction of diaphragm wall and rock anchors also needed special equipment and machinery. Such a work could not be entrusted to sub-contractors doing normal building works and selection of the firms who were specialized and experienced in such foundation systems was, therefore, considered essential. Accordingly, the need was felt for pre-qualification of sub-contractors exclusively for the foundation system. Similarly, good agencies of proven ability for water proofing treatment considered in pre-qualification. However, to avoid any problem of coordination amongst various sub-contractors, it was decided to go in for tender, both for the foundation system as well as its water proofing treatment with the provision that the main sub-contractor would engage an agency out of the pre-qualified firms for water proofing treatment.

8.1.7.1 *Purchasing Data*

The sufficient data was made available to the subcontractor with necessary details of soil investigation, availability of site, security restrictions, traffic restrictions, avoidance of disturbance of various items, conditions of dewatering, etc. were supplied as part of the tender document. Both in respect of foundation work as well as superstructure work, in order to make the relevant data free from any ambiguity, two pre-bid conferences were held in each case with the pre-qualified firms and the clarifications, suggestions as deliberated, were incorporated in the purchasing data.

8.1.7.2 *Verification of Purchased Product*

The basic materials, viz. cement and steel conforming to specifications were procured through Central Stores of the CPWD from reputed firms to ensure quality. Tests were being regularly conducted on other materials immediately on their arrival at the site. Materials were stored in proper sheds to prevent them from any damage. Any sub-standard materials brought at site were immediate identified, reported to sub-contractor and their removal ensured.

For various items of the work executed by the sub-contractors, regular inspections were carried out by the departmental officers at various stages to verify that the same conforms to the specifications.

8.1.8 Product Identification Traceability

The building superstructure was divided in different blocks with expansion joints and these blocks were named after their functional use. Cartesian coordinate system was being followed to identify different areas of the site. Columns were identified by grid line intersections. Each panel of diaphragm wall was numbered. Each rock anchor had its own identification number. Cube tested register indicated the location of the concrete laid and was represented by the particular cube sample of the concrete. Date of casting and the component of the work where cement was used, was also indicated in the cement register.

All the relevant data regarding various operations in respect of a particular area or activity were properly documented quoting the aforesaid identification numbers.

8.1.9 Process Control

Procedures for various processes with applicable standards, suitable equipment and their maintenance, suitable working environment, etc. were prepared including the specifications stipulating verification of workmanship based on appropriate samples and required tests. The detailed work instructions for the foundation system for the new and unconventional items of diaphragm wall, curtain grouting, and rock anchors and also that of water proofing treatment were available therein. Job instruction/check lists wherever required were issued to the field engineers for various activities for process control.

Due consideration was given to deploying trained personnel for special processes and it was ensured that only those who had desired level of competence were assigned such construction processes.

8.1.10. Inspection and Testing

Inspection and test plans had been prepared to document all the testing and inspection requirements relevant to a particular operation or element of work such as receipt of incoming materials, during construction and final completed works.

The materials were received and properly stored in identified areas at site. These were not used for the construction works until test and inspection had been carried out.

Inspection or tests were carried out at all stages of construction to verify conformity with standards, specifications and drawings. The same was also checked frequently by the senior officers. Until the same had been done, next stage of construction was not undertaken.

No construction work was treated as complete unless all the activities as specified in the associated data and documentation for the same were made available. Deficiencies or deviations noticed during inspection and testing were rectified, which on rectification were again inspected.

Executive engineers were responsible for establishing and maintaining records in proper formats to provide evidence that constructed works met the acceptance criteria as laid down in the purchasing data. In case of failure of a product to pass any inspection and/or test, the procedure for control of non-conforming products was made applicable.

8.1.11 Control of Inspection, Measuring and Test Equipment

Equipment register was maintained to identify the type of equipment, the method to be used for its calibration, and the frequency. All the measuring and test pieces of equipment were inspected periodically. They were got tested and calibrated at specified intervals from the approved laboratories/ agencies. It was also ensured that no damage or loss in calibration had taken place during handling, transportation, storage, etc. of such equipment. Steps were also taken to prevent tampering of the instruments after they had been calibrated.

8.1.12 Inspection and Test Status

The constructed work could have one of the following inspection/test status: (a) Work was yet to be inspected; (b) work had been inspected and passed; (c) work had been inspected and rejected; (d) the constructed work on inspection, if not found acceptable, could be repaired, rectified or utilized under concession after approval of the designated authority, superintending engineer being the authority in the instant case.

The inspection the test status could be seen from the site order book, inspection register and checklists. Cross markings at site were given for rejected materials/works. Compliance was indicated against the above entries. Only the accepted materials/works were measured and entered in the measurement books.

8.1.13 Control of Non-conforming Product

All the materials works which did not conform to specified requirements were prevented from use or installation. All construction works not conforming to standards and specifications were immediately identified with suitable marking at site and simultaneously suitable entries were made in the site order book. A separate report was prepared for such non-conforming works indicating their identification, quantum affected, details of fault, defect or non-conformity and the stage at which the same was discovered. This report was referred to a Correction and Preventive Action Committee which suggested about its disposition in either of the following ways.

   i. Dismantle and reconstruct to meet the specifications and standards
  ii. Accept without dismantling but with or without repairs on concession.
 iii. Put to alternative use with suitable modification, or
 iv. Reject or scrap

All such reviews were recorded and documented. Appropriate steps were taken to prevent recurrence of non-conforming works.

8.1.14 Corrective and Preventive Action

In the instance of non-conformity case, a Corrective and Preventive Action Committee had been constituted with superintending engineer as its chairman and executive engineers as its members to investigate and determine the potential causes for non-conforming works. This committee had been delegated the responsibility and authority to look into various aspects like the design, specifications machinery and equipment, materials operating and supervisory staff, process control or even complaints from subcontractor/client for analyzing the problem.

Depending upon the type of problem which was encountered during the construction which resulted in sub standards/non conforming construction, the above committee was to evaluate the corrective measures. Corrective action report prepared by the committee was then communicated to the concerned personnel for suitable action. It was ensured that with the above solutions of the problem, the non-conformity had been minimized to the required extent or eliminated.

8.1.15 Handling, Storage, Packaging, Preservation and Delivery

Its purpose was to prevent damage and to ensure safety of materials during construction and also to ensure proper protection and handover as built construction work. Detailed instructions for handling, storage and maintenance with suitable storage facility were mentioned in the documented data.

On receipt of a materials from vendors/suppliers/stores, the same was properly accounted for in the relevant registers and ledgers indicating the name of the supplier, purchase order number, quantity received, date of receipt, supplier's inspection and test certifications, etc. It was ensured that no damage deterioration/contamination of materials or fabricated articles/works occurred during their handling. Detailed instructions were available for handling, storage and issue of materials and products applying the principle of first-in-first out wherever applicable. Suitable storage space was ensured for various materials. All precautions were taken for proper handling and storage of various semi-finished and finished components of works.

8.1.16 Control of Quality Records

Quality records were retained minimum up to completion of project and retention period could be extended till settlement of disputes, claims and arbitration.

8.1.17 Internal Quality Audits

The internal quality audit was undertaken by independent persons or bodies having no direct interest or responsibility in respect of the work being examined. Audits were performed in accordance with written procedures or checklists.

8.1.18 Training

The training needs were established to keep the staff abreast with latest technological developments in construction field and also to continuously improve upon their performance. The training courses were organized through the chief engineer (training) in charge of the CPWD training institute. Circulation of periodical journals technical books and screening of video films on technical subjects were also resorted to train and motivate the staff.

8.1.19 Statistical Techniques

The statistical techniques were identified for construction materials.

8.1.20 Concluding Remark

Companies belonging to the manufacturing sector had gone successfully for ISO 9000 certification. Implementation of ISO 9002 standard in Parliament Library Building Project indicated that standard is equally applicable to construction industry. The project team meticulous planning and systematic approach, made it possible to evolve a quality system conforming to the provisions of ISO 9002 standards for execution of a building project. This required a lot of documentation and record keeping which was found to be in the larger interest of all the stakeholders, involved in the building project.

8.2 CASE STUDY II: IMPLEMENTATION OF ISO 9000 STANDARD FOR FLYOVERS PROJECT IN PWD, DELHI

8.2.1 Principal Requirement of Quality System

8.2.1.1 It has been universally acknowledged and accepted that quality can be maintained and assured through continuous surveillance that has to be mounted by the client by prescribing norms and procedures to ascertain the conformity of the construction to the predetermined specifications. To achieve the desired goal, quality management has to be adopted to determine the quality policy and to implement it by such means as quality planning, quality control and quality assurance with its quality system. The quality of construction is ensured and maintained by following a documented "Quality Plan" which in turn is defined as the operational techniques of controlling quality. Quality assurance is an aspect of the quality which encompasses all the planned actions that are needed to provide sufficient confidence that the construction will meet the needs/requirements. Quality assurance is a system of planning, organizing and controlling human skills to assure quality. Quality assurance plan sets out the planned actions required for quality assurance. Quality assurance is achieved by continuous monitoring and verification of the QA activity to ensure that it is being followed and this is known as quality surveillance. Quality policy is a formality documented statement of management's intensions and directions as regards quality.

8.2.1.2 Contractor's policy is a commitment of the agency to achieve and sustain quality of construction to meet client's stated needs which are contractually or otherwise required. It is achieved by adopting quality assurance system by resorting to documentation through quality policy manual, procedure manual, quality plan, quality control and method statements related to construction techniques/methods of construction/methodology of sequence of operations, construction programme taking into account release of approved construction plans and other necessary inputs by the client, well before the start of any component of work. Client's policy is a commitment of the client to obtain construction quality to meet his stated or implied needs. To achieve the stated targets and desired results, client has to prescribe tests and procedure and levels of quality checks.

8.2.1.3 The quality assurance of a construction project, starts from the project preparation and formulation stage which includes pre-planning, planning and design, preparation of detailed drawings and estimates, tender documents, selection of competent agency for execution of the work and so on. The client may have either in-house capacity to carry-out all these activities or some of these activities can be out-sourced if sufficient in-house capacity or know-how is not available. The client must have a clear vision of the requirements which have to be kept in focus while formulating a project after its need is established. These requirements include: (i) Proper planning and requiring for speed of construction, (ii) standard and specifications to which the work has to be done, (iii) methodology of construction to be adopted suiting the particular project by taking into account the site conditions, (iv) cost effectiveness of the chosen construction technology, and environmental impact aspects.

The above mentioned factors have to be given due importance in order that a proper and relevant project formulation takes place. The clients are increasingly outsourcing the project, formulations for which selection of the specialists is of paramount importance. The selection of proper planner or designer must take into account the qualification and experience of the key personnel of the organizations that are invited to bid for the project formulation.

8.2.1.4 The client's capacity of specifying the detailed work requirements and to judge the output given by the consultants cannot be over emphasized. It is needless to say that there is no substitute to a well-educated and a vigilant client. While the selection of the consultant is important for proper project formulation including cost estimate and tender documents, selection of contractor for executing the project is equally relevant. Therefore, laying down stringent criteria for the selection is essential. This can be done either through the process of pre-qualification or post-qualification; former being preferred. A pre-bid conference is essential to remove any grey areas before selection of contractor for executing the job. Although the selection of consultant and contractor is not directly related to project quality management, yet they are considered as the two most important factors in ensuring project quality management.

8.2.1.5 The plans and designs prepared by the consultant and the shop drawings prepared by the contractor, should be subjected to check and review by the consultant when high QA level is specified. When extra high QA level is prescribed, the review by an external organization (also known as proof consultant) is essential. The selection of the proof consultant is a critical activity and extreme care and caution is called for in this process. The experience of the key personnel of the proof consultant should be either equal to or higher than that of the consultant. In order to monitor and enforce project quality management, preparation of QA manual is required. The QA requirements may be covered in the tender documents or the contractor may submit the same in addition, and get approval of the QA manual from the client, defining all activities with minimum two-level control or three-level control of QA which may be normal, high or extra high. It is recommended that the level of quality assurance for projects like bridges and flyovers should be extra high (Q-4 level as defined in IRC:SP:47.

8.2.2 Features of Implementation of ISO 9000 Standards in Delhi Flyover Project

8.2.2.1 In 1993–2000, a number of important flyover projects were planned and executed by PWD Delhi. These projects include:

    i. Four level flyover cum underpass at Punjabi Bagh.

    ii. Segmental flyovers with approaches of Reinforced Earth at Moti Bagh and Safdarjung Enclave on Ring Road as also Savitri Cinema and Nehru Place on Outer Ring Road.

    iii. Flyovers with structural steel and Reinforced Earth approaches near Kirti Nagar and across Khelgaun Marg on the Ring Road.

    iv. Flyovers at Dhola Kuan and AIIMS on Ring Road.

    v. Underpass near South Extension on Ring Road.

8.2.2.2 In order to achieve quality standards of the highest order, it was decided to take ISO quality system certification for these projects independently for each contract. According to requirements of ISO standard management responsibility was decided, quality system documentation was done, design and data controls were exercised, traceability of products ascertained, and process controls were done. To ensure proper implementation, ISO system consultants were appointed, engineers of PWD, and contractors given training and processes established as also recorded. The specific case study of segmented flyovers on Ring Road and Outer Ring Road is given here.

8.2.2.3 The quality assurance manual was developed for these flyovers constructed by using reinforced earth technology for the solid approaches and precast, pre-stressed concrete segmental technology for the super stricture. In order to follow ISO standards, PWD Delhi, for the first time in the country, followed provisions of IRC:SP:47 (Guidelines on Quality Systems for Road Bridges), published by Indian Roads Congress. This manual contained quality assurance plan for each of the major components of construction from foundation to superstructure. It also contained check lists and approved method statements for carrying out all the cardinal activities related to the various components of the flyovers. The approved method statements ensured that the chances of a mistake during the execution of the related activity or item of work were minimized as the supervisors of PWD and contractor knew exactly as to what was required to be done. As a part of the said manual, proper forms were developed and approved to document the identification of source of materials like cement, steel, coarse aggregate, fine aggregate, bridge bearings and so on with the aim of keeping their traceability on record. The source and properties of the various materials used were recorded. The test results conducted on the different raw materials and finished products had also been documented systematically in the proformas available in the IRC SP 47, or evolved separately so as to ensure the accountability of the concerned personnel of the PWD and contractor.

8.2.2.4 Since a foolproof system was evolved to record various operations like production of concrete and its transportation and placement, the work supervisors were vigilant enough to avoid any possible failure. Any non conformance to procedures could be easily pin pointed and attributed to particular personnel connected with the execution of the related activity or item of work. As per para 3.3 of IRC:SP:47, three/four level controls were specified in so far as quality of different materials was concerned. These are as follows:

    i. The names of manufactures/suppliers for materials like steel and cement were specified in the tender documents based on the reputation of such manufactures/suppliers. Similarly, the brands and process of systems like pre-stressed concrete and manufacturers of specialized items like bearings, and expansion joints, epoxy compounds were specified in the tender documents. The agency was required to submit the test report of the manufacturer/supplier of the materials at the time of submitting the proposal to approve them for the work.

ii. Independent testing from one of the approved test houses was also resorted to, in order to doubly check the quality of the materials.

iii. In addition to (i) and (ii) above, the testing of the materials at different intervals was also done during execution of the work.

iv. The in-house testing facilities, set up by the contrctor were also availed for testing the quality of materials.

The quality assurance manual finalized for the specific work was generalized so that it could be made use of with case specific variations in other situations like cast-in-situ construction and other technologies. This manual included by and large typical quality assurance plan for design and construction of the project according to IRC:SP:47. The quality assurance manual was sub-divided into ten chapters. Each chapter deals with one distinct component of the Project Quality Management in sufficient details. The manual was published by CPWD.

8.2.3 Advantages and Conclusions

8.2.3.1 The adoption of quality management during the different stages of planning and execution of the project had brought about a revolutionary change in the thinking of the engineers and managers associated with the projects. As each activity forming a component of the project got defined and recorded clearly, the possibility of any aspect being missed during execution got eliminated. This was achieved by including checklists for each activity or item of work. A scrupulous adherence to the checklists would avoid any pit fall.

8.2.3.2 Furthermore, as the sources of various materials and products incorporated in the work were traceable, their manufacturers/suppliers become vigilant with regard to the conformity of the specifications of such materials to the standards mentioned in tender documents or plans. It is also relevant to mention that the supervisory staff became vigilant during the processes of production or laying as they could be traced in case the quality of work was subsequently found to be below standards. In a nut shell a properly prepared and executed project quality plan leaves nothing either to chance or to the personal whims and fancies with the result that the end product is strong and durable and fulfills the criteria enunciated at the planning and conceptual stage of the project. A proper documentation of the Project Quality Management is bound to be beneficial both for the agency as well as the client. It is more in the interest of the client that the working should be adapted to a well thought out quality plan which would ensure speedy and timely execution ensuring adherence to the laid down norms, standards and specifications during the execution of the project.

8.2.4 IRC SP-47, 1947–1948, was based on quality systems evolved by ISO 9000 standards. This document was drafted by very eminent highway engineers of the country. Shri NV Merani who was the convener and S/S S.G. Joglakar, AG Barkar, SA Reddy, PV Manjure and Shitla Sharan, were members. Therefore, it was decided that for flyover projects of PWD, ISO 9000 procedures, as detailed in this publication would be followed.

Both PWD engineers and contractor's engineers took training from ISO standard certifying agencies and, thereafter, implemented ISO Standard for these projects. The ISO standard certifying agencies were also engaged to evolve systems and ensure the implementation.

The necessary documents required or ISO 9002 certification were prepared. CPWD specifications were very useful in preparation of these documents as quality aspects like frequency of testing, method of testing for materials, etc. were covered in CPWD specifications and were being followed for works. The PWD Delhi was awarded ISO 9002 certification for establishing and maintaining a quality system for construction of these flyover projects.

8.3 CASE STUDY III—ISO STANDARDS CERTIFICATION FOR INTGRATED MANAGEMENT SYSTEM (IMS) OF M/S AHLUWALIA CONTRACTS (INDIA) LIMITED, NEW DELHI

8.3.1. About Company's Integrated Management System

- Ahluwalia Contracts (India) Limited (ACIL) is a deemed Public Limited Company under the Company Act. 1956. The company has more than 500 engineers and total staff of approx. 1700, its head office is located in New Delhi. ACIL offers construction and related services on turnkey basis. ACIL is the first company in India which took certification for ISO 9001 Quality Management System (QMS), ISO 14001 Environmental Management System (EMS) as well as for BS OHSAS 18001 standard related to health and safety. Presently, ACIL is committed to the establishment and maintenance of Quality, Health, Safety and Environment (QSHE) Management System in accordance to the requirements of ISO 9001:2015, ISO 14001:2015 and ISO 45001:2018 standards. Certification for ISO 45001 Occupational Health and Safety (OH and S) Management System is under progress while certifications for ISO 9001:2015 QMS and ISO 14001:2015 EMS were attained in the year 2017. The IMS implementation approach has enhanced staff and worker involvement in each activity.

- Top management has nominated a management coordinator (MC) for reviewing and updation of the QSHE manual. Major changes to the manual affecting the QHSE are approved by the Management Review Committee. The company has documented all the processes needed to ensure establishment, implementation, maintenance and continual improvement of QEHS management systems. The processes are carefully mapped and analyzed for their effective implementation.

- **Quality, Safety, Health and Environment Policy (QSHE Policy)**
  We, at Ahluwalia Contracts (India) Limited, are committed to continually improve our product quality, environment, health and safety performance, by applying management practices:

a. Meeting and exceeding customer requirements by giving them on time delivery and product consistency.

b. Commitment to comply with all legislative and other requirements relating to environment, health and safety; commitment to prevent incidents and environmental pollution.

c. Continual up-gradation of processes and technologies with emphasis on improving product quality, environment, health and safety.

d. Operational excellence by optimization of cost, improvement in productivity and good construction practices.

e. Enhancement of people capability by providing education and training.

8.3.2. Context of the Organization

8.3.2.1 Context

The company has determined external and internal issues that are relevant to its purpose and its strategic direction and that affect its ability to serve its purpose and achieve the intended results of its integrated management system. Some of the examples of external context are: Statutory and regulatory requirements; technology used in processing; suppliers of construction materials, finishing materials and consumables; subcontractors for construction, design and architectural consultants; second party and third party inspection agencies; testing laboratories; local communities and related issues; state level or national level issues/strikes leading to material non-availability; natural calamities (flood/heavy rain/earthquake/accidents, etc.); logistics issues related to movement of construction materials and other items; external resources like power/water, etc. Similarly, some of the examples of internal issues that impact on company's functioning are: Resources (capital, time, people, process, systems technologies); roles and accountabilities of human resources; knowledge level of employees; policies, objectives of the company; information systems, information flows and decision-making processes.

8.3.2.2 Needs and Expectations of Interested Parties

The interested parties were identified as per management system standards, and their related to needs and expectation were identified. For example, subcontractors require timely payment of completed activities and suppliers of materials require timely inspection and approval of materials.

8.3.3. Leadership and Commitment

Processes for consultation and participation of workers at all applicable levels have been implemented for management system involvement and effectiveness.

8.3.4. Planning for the Quality, Environment, Health and Safety Management System

- Risks and opportunities are determined considering the context of the organization, scope, needs and requirements of the interested parties. Unacceptable hazards have been identified through a process of hazard identification and risk analysis (HIRA) and documented. Significant environmental aspects and unacceptable risks/hazards are communicated through the process of "consultation and communication". Hazards identification should proactively identify any sources or situations, arising from an organization's activities, with potential for work-related injury and ill health.

- System is established and implemented to assess risks from the identified hazards, while considering the effectiveness of existing controls.
- System is established to implement and maintain processes to assess:
  a. Opportunities to enhance performance, while considering, planned changes to the organization, policies, processes and activities.
  b. Opportunities to adapt work, work organization and work environment to workers.
- Integrated objectives have been established and documented at relevant functions, levels and processes, considering the significant environmental aspects, unacceptable hazards and applicable compliances obligations, and considering its risks and opportunities.
- Change to the QEHS management system are carried out in a planned and systematic manner as per the process of "Change Management".

8.3.5 Resources and Organizational Knowledge

- Resource planning is carried out periodically to determine and provide the resources needed for the QEHS management system.
- Infrastructure is maintained for the operation of its processes to achieve conformity of products and services.
- The resources needed to ensure valid and reliable monitoring and measuring results are maintained.
- Procedures defining appropriate corrective action as necessary if the validity of previous measurement results has been adversely affected when an instrument is found to be defective during its planned verification or calibration, or during its use.
- Documentation of knowledge necessary for the operation of its processes and to achieve conformity of products and services has been accomplished.

8.3.6. Competence

The organization has determined the necessary competence of persons doing work under its control that affects its QEHS performance to:

- Ensure that these persons are competent on the basis of appropriate education, training or experience.
- Where necessary actions are taken to acquire the necessary competence, and evaluate the effectiveness of the actions taken.
- Evidence of competence is maintained as "Skill Matrix".

8.3.7. Awareness

- All persons working under the organization's control are made aware of—the QEHS policy, relevant QEHS objectives.

- The significant environmental aspects and related actual or potential environmental impacts associated with their work.
- Their contribution to the effectiveness of the QEHS management system, including the benefits of enhanced QEHS performance.
- The implications of not conforming with QEHS management system requirements, including not fulfilling the organization's compliance obligations.

8.3.8 Communication, participation and consultation

- Process for "communication and consultation" has been documented for internal and external communication relevant to the environmental, health and safety management system.
- Internally communicate information relevant to the QEHS management system is communicated among the various levels and functions of the organizations, including changes to the QEHS management system.
- The organization externally communicates information relevant to the QEHS management system, as defined by the management and as required by compliance obligations.

8.3.9 Operational Planning and Control

- Processes are defined and all activities planned and implemented under controlled conditions to meet requirements for the provisions of products and services and environment, health and safety management systems.
- Documented information is maintained and retained as necessary to ensure that the processes have been carried out as planned and to demonstrate conformity of products and services.
- Systems are established to control the procurement of products and services in order to ensure their conformity to its integrated management system.

8.3.10 Monitoring, Measurements, Analysis and Evaluation

- Processes have been defined and documented to evaluate effectiveness of integrated management system.
- Environmental performance and the effectiveness of the environmental management system is evaluated.
- Customer perceptions are monitored to assess the degree to which requirements have been met. Process has been defined to obtain information relating to customer views and opinions of the organization and its products and services.
- Documented process is established to implement and maintain and evaluate fulfillment of its compliance obligations by:
    a. Determining the frequency and method for the evaluation of compliance.
    b. Evaluating compliance and act if needed.
    c. Maintaining knowledge and understanding of compliance status.

8.3.11 Internal Audit

- Process has been established to conduct internal audits at planned intervals to provide information on whether the QEHS management system conforms to:

  a. The organization's own requirements for its QEHS management system.

  b. The requirements of this International Standard.

  c. Is effectively implemented and maintained.

- The process for internal audits defines, audits programs including the frequency, method, responsibilities, planning requirements and reporting, which shall take into consideration the QEHS objectives, the importance of the processes concerned, customer feedback, changes impacting on the organization, and the results of previous audits.

- A pool of internal auditors for QEHS audit have been developed by training on ISO 9001:2015, ISO 14001:2015 and ISO 45001:2018 standards. The scope of audit generally remains limited to company's QEHS audit. However, sometime at projects the scope of audit changes due to specific requirements of client.

8.3.12 Management Review

- Process of management reviews has been established to review the organization's QEHS management system, at planned intervals, to ensure its continuing suitability, and effectiveness.

- **Improvements**

  Opportunities for improvement are selected and implemented to meet customer requirements and enhance customer satisfaction.

These include:

  a. Improving processes to prevent nonconformities.

  b. Improving products and services to meet known and predicted requirements.

  c. Improving QEHS management system results.

8.3.13 Incident, Nonconformity and Corrective Action

- Documented processes are established to, deal with nonconformity and take action to control and correct it and deal with the consequences.

- **Incident Investigation**

  Procedure is maintained to record, investigate and analyze incidents. The results of incident investigations shall be documented and maintained.

- **Continual Improvement**

  Continual improvement of the QEHS management system is part of policy and strategy of the organization.

8.3.14 Conclusion

ISO Management Systems enabled ACIL to create an integrated approach towards the core business functions, building more customer focused processes and improve the processes continuously to ensure company's performance consistent with the changing demands of business. It also helped to increase the employee's morale by ensuring that the roles and responsibilities are clearly marked to create a better accountability and clarity of their roles which helped in creating a better work environment.

8.4 CASE STUDY IV-ISO 9001:2015 QMS TRANSITION OF INTERCONTINENTAL CONSULTANTS AND TECHNOCRATS PVT. LTD., NEW DELHI

8.4.1 Introduction

ISO is always considered as a vital element of business planning as well as a useful management tool in ICT Pvt. Ltd, a leading international consultancy company. The company attained its first ISO 9001 certification in September 2000, which was based on quality assurance model confirming to ISO 9001:1994 Standard. Progressively, the ISO 9001 standard has kept on evolving by responding to the changes in need of the business and market. Simultaneously ICT Pvt. Ltd. continued to advance ISO 9001 certification conforming to Quality Management System (QMS) requirements for the versions as per 2000, 2008 and 2015 standards. This case study shows how ICT Pvt. Ltd. started and finished the QMS transition to ISO 9001:2015, and the challenges faced during the re-certification.

The maiden discussion on ISO 9001:2015 standard was held in ICT's management review committee (MRC) meeting dt. 29th December 2015 and the necessity to transition to the new requirements was established amongst all committee members. The benefits of the new system over the existing one were discussed at length to overcome the "Resistance to Change" during implementation.

The practice of ISO 9001:2015 quality system in the organization brings with it a set of procedures that define the method to be followed for doing a task. This brings in uniformity and continuity in the team across disciplines and across time-lines, thus helping to align with strategy and vision of the management.

8.4.2 Brief About the Company

Intercontinental Consultants and Technocrats (ICT) Pvt. Ltd., is a premier engineering consultancy firm, providing comprehensive professional services in all facets of infrastructure projects including planning, designing and construction supervision. Founded in 1987, ICT has engineered a large number of infrastructural development projects with a spirit of innovation, dedication to quality and skills. The sectors in which ICT has been offering consultancy services are highways, bridges, aviation, urban infrastructure development, traffic and transportation, metro structures, railways and metro allied services, water resources, water supply and public health engineering, ITS: (Intelligent Transport Solutions), power, architecture, surveys and GIS mapping/analysis and social and environmental sciences.

The company under its Chairman and Managing Director (CMD), Chief Operating Officer (COO) and its Presidents, has set-up an organization with a defined hierarchy and interfaces. The QMS of ICT focuses on the achievement of results, through quantified quality objectives, to satisfy the needs, expectations and requirements of its clients. Also, the system provides a framework for continual improvement of its effectiveness aimed at enhancing client satisfaction.

Top management setup is given in Fig. 8.1

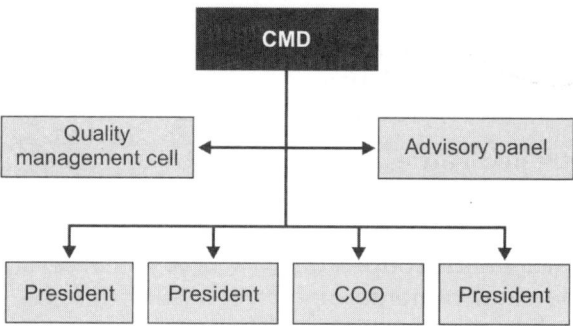

Fig. 8.1: Top management

8.4.3 Developing a Transition Plan

Considering the importance of the transition for the company, the first step undertaken was to determine how much work was ahead, so quality management cell (QMC) team conducted a gap analysis to determine to what level the company was already compliant with the standard, and what needed to be done to achieve full compliance. The gap analysis showed that there were many documents to be updated in order to be aligned with ISO 9001:2015, and some additional processes where needed to be implemented, such as determining context of the organization and addressing risks and opportunities.

According to the gap analysis findings, a transition team was formed that included heads of each department and responsible persons for each process in the company. Each was given their role in the transition project and assignments for different tasks. The tasks included writing new documents, updating old ones, and updating and developing new processes.

Activities related to clauses 4, 5, 6, and 7 of ISO 9001:2015 were assigned to Head-QMC, while the rest of the team dealt with the other clauses of the standard related to their processes. The adequacy of Quality Policy and Vision Statement were discussed in Head-QMC meeting and revised policy and vision statement where implemented with effect from 4th September 2017.

8.4.4 Transition Activities

During preparation for ISO 9001:2015 re-certification, the trainings were conducted for awareness cum sensitization, implementation and for internal audits. Some problems were encountered and effectively resolved: (a) Resistance to change, (b) turnover of staff and internal auditors, (c) delay in up-gradation of the procedures and (d) auditee's non-cooperation with auditors.

The Head-QMC presented to the top management and the transition team methodology for ISO 9001:2015 transition and responsibilities of all concerned as detailed below.

1. Determine Context of the Organization

Determining context of the organization was a new requirement and the Head-QMC decided to call the top management and the transition team to participate in determining the context.

It was decided that each member of the transition team would conduct the analysis independently, and the results would be merged into a single document afterward. In Management Review Committee meeting the issues relevant to the QMS were identified and discussed.

2. List All Interested Parties

As a part of determining context of the organization, the same team was involved in identifying interested parties. They focused only on the relevant interested parties and grouped them according to their requirements. In MRC meeting, the requirements of interested parties relevant to QMS were identified and discussed.

3. Determining the Scope of the QMS

The scope of QMS was re-assessed in light of the external and internal issues and the requirements of relevant interested parties. The company decided not to change the scope of QMS for the organization.

4. Demonstrate Leadership

In MRC meeting the ISO 9001:2015 requirements pertaining to leadership and commitment of top management for QMS were explained and discussed for its effectiveness. Head-QMC also conducted an interactive session with all functional HoDs to clarify the leadership requirements of ISO 9001:2015 standard. Most of these requirements were to be met through actions rather than documentation.

Having a chance to implement the standard by themselves, each HoD got a better understanding of the standard and, therefore, got better insight on how the standard could help them in their line of work. The quality policy was updated, through a joint effort from the top management and the transition team, to meet all requirements of the standard. After the final approval of the CMD, it was communicated to interested parties through website.

5. Align QMS Objectives with the Company's Strategy

The top management decided to frame short and real functional objectives for the QMS. Involvement of the top management in defining quality objectives ensured that they were aligned with the company's strategy.

The new version of the standard required a new approach to defining objectives; besides defining SMART (specific, measurable, achievable, realistic and time bounded) objectives, the company also defined plans for achieving these objectives. The plan included defining the responsible person, a set of activities that would lead to achievement of the objective, necessary resources, and deadlines.

6. Assess Risks and Opportunities

Considering that all relevant people in the company participated in determining context of the organization, the risks and opportunities became clear from the start. The transition team decided not to define methodology for addressing risks and opportunities, but to arrange a brainstorming session with the top and middle management and discuss the risks and opportunities in the company as a whole, and at the process level. The team decided not to write a procedure for addressing risk and opportunities, but rather to only create a risk register which contained information about the risks and opportunities, risk owners, and details about actions to be taken to address the risks and opportunities.

As an input for the identification of risks and opportunities, the previously determined context of the organization, records about previous nonconformities and customer complaints, and customer satisfaction survey results were used. This allowed to identify all risks and opportunities relevant to the QMS and to distinguish them by their significance to the company. Afterwards the significant risks and opportunities were defined, plans for addressing risks and opportunities were made in a manner similar to their plan for achieving quality objectives.

7. Control Documented Information

The company had one procedure for document control, and another one for record control. These procedures were quite long with robust document coding system for approval and withdrawal. The rules for document and record control were too complicated and often led to noncompliance and nonconformities in the past. To meet the requirements of the new standard, the two existing procedures were merged into one in more simplified manner.

8. Operational Control

The existing QMS was created according to the "document all you do" approach and the key processes were explained with documented procedures and work instructions. Soon, it was realized that replicating the same approach would take too much time for the processes to be introduced into the scope of the QMS. The only way to resolve this issue was to try a different approach.

The Head-QMC decided to look into the standard and see what exactly the standard required. He decided to first determine the requirements for the products provided consistent with information already existing in the QMS documentation. Then, with the help of relevant process owners, it was decided to define the criteria for each process, meaning defining the activities, monitoring and measuring and required resources. Considering the level of competence and experience of the employees working in each process, the Head-QMC and the process owner decided what documented procedures and work instructions were really needed, and what could be discarded.

Because most of the employees had worked in the company for a long time, they didnot really need too many documents. Most of the work instructions turned out to be redundant, because the employees had sufficient levels of experience and knowledge to manage their responsibilities without such written instructions. On the other hand, training was regularly imparted to the employees to ensure that they were competent for the work. The Head-QMC and the process owners observed the processes and determined which activities were too complex, and where the non conformities were most likely to occur, and decided to have work instructions for suchactivities. In the revised QMS, the processes were documented by defining the activities, resources, responsibilities, monitoring and measuring activities, and reference documents for each process.

9. Review the Design and Development Process

The design and development process needed significant updates to become aligned with ISO 9001:2015 requirements. The design and development process were too complicated and in some cases lacked the requisite controls. The process owner readily accepted the task of updating the process. It was started from planning the design and development and defining all inputs for the planning, from considering the nature, duration, and complexity of the design and development activities to defining what documented information would be needed to demonstrate that design and development requirements had been met.

The design and development processes were often interrupted by frequent changes in requirements, so it was decided to document in detail how the changes would be identified, reviewed, authorized, and executed, including activities to avoid adverse effects from the changes.

10. Control of External Providers

The company had external providers such as suppliers and outsourcing agencies. The purchasing process was decided to be systematized and update the requirements for externally provided processes, products, and services in order to ensure conformance to the specified requirements. The controls to be applied to suppliers and outsourcing agencies were determined to ensure that the products and services were compliant with defined requirements.

11. Performance Evaluation

The CMD always wanted to have a system of monitoring and measuring that could provide information on the overall condition of the company and process performance. This was achieved through internal audits and management reviews, including process performance assessment through performance indicators.

The revised QMS system slowly started to fall into place, and after monitoring and measuring key performance indicators, customer satisfaction, and performance of suppliers and subcontractors, the inputs for effective management review started to become evident.

8.4.5 Certification Audit

The ISO 9001:2015 QMS was implemented in the company on 28th November 2017, subsequently two internal audits were conducted to understand the status before re-certification audit. The results of the internal audit and management review were found satisfactory in meeting the requirements of ISO 9001:2015.

The certification audit held on 21st and 22nd May 2018 by DNV-GL, the auditor's team did not have any major remarks on the system and after resolving a minor non conformity and few observations, the company got the certification. The external auditors conveyed, during the closing meeting, that implementation of ISO 9001:2015 QMS was of a high order.

8.4.6 Conclusion

On 14th June 2018 the ICT Pvt. Ltd. became ISO 9001:2015 certified organization. The transition was a success and it brought many improvements to the existing QMS and benefits to improve business performance. ISO 9001:2015 is considered as the performance driven management system for enhancement of quality, customer satisfaction and improvement.

The changes in quality management principles, QMS process model and risk and opportunity management concept were considered helpful in inculcating the principles of quality in the organization. The benefits accrued by the organization are: Competitive edge over other consultants, enhanced recognition in the national and international markets, defined responsibility and accountability at all levels, proactive pursuit of client satisfaction by all concerned and hence improved client satisfaction, technology upgradation, improved communication amongst divisions, better working environment, improved feedback system and a motivated workforce.

Index